Chemical Consequences

Chemical Consequences

Environmental Mutagens, Scientist Activism, and the Rise of Genetic Toxicology

SCOTT FRICKEL

RUTGERS UNIVERSITY PRESS

NEW BRUNSWICK, NEW JERSEY, AND LONDON

LIBRARY OF CONGRESS CATALOGING-IN-PUBLICATION DATA

Frickel, Scott.
    Chemical consequences : environmental mutagens, scientist activism, and the
rise of genetic toxicology / Scott Frickel.
        p.   cm.
    Includes bibliographical references and index.
    ISBN 0–8135-3412-7 (hardcover : alk. paper) — ISBN 0-8135-3413-5 (pbk : alk.
paper)
    1. Genetic toxicology.   2. Chemical mutagenesis.   3. Mutagens.   I. Title.
    RA1224.3.F75 2004
    616'.042—dc22   2003019796

British Cataloging-in-Publication information for this book is available from the
British Library.

Manufactured in the United States of America

For Beth

CONTENTS

FIGURES AND TABLES

FIGURES

TABLES

ACKNOWLEDGMENTS

This book began its life in courses and research begun in Madison, Wisconsin; was written in Philadelphia; and was completed in New Orleans. It has benefited from the direct and indirect influence of numerous people along the way. I owe intellectual debts to many, but the work of four scholars stands out. Robert Kohler's studies of biochemistry, genetics, and science patronage have done much to shape my thinking about the political economies of discipline and disciplinary practice. Bruno Latour's *Science in Action* first drew me to science and technology studies, and I have spent much of the past decade entangled in actor networks. I had the great fortune in graduate school to take two courses with environmental historian William Cronon, where I learned important lessons about the place of nature in history and about the place of history in the narratives we construct about nature. Finally, John Walton's (1992) short essay titled "Making the Theoretical Case" helped me understand how cases are made and why mine mattered.

Conversations in 1996 with Mayer Zald, Michael Kennedy, Howard Kimmeldorf, Dieter Rucht, and John McCarthy during a visiting fellowship at the University of Michigan–Ann Arbor gave form to my embryonic ideas about the relationship between science and social movements. My dissertation committee members—Charles Camic, Fred Buttel, Warren Hagstrom, Lynn Nyhart, and Pam Oliver—peppered me with difficult questions and thoughtful advice when I sought it from them and left me alone when I didn't. A discussion with Daniel Kleinman twelve years ago about Alvin Gouldner's "new class" theory has since grown into an ongoing and far-ranging conversation about the politics of knowledge that never disappoints. I remain indebted to many former graduate school colleagues for sociological insight, to be sure, but even more for their camaraderie, courage, and trust: Stuart Eimer, Jim Elliot, Beth Fussell, Neil Gross, Jonathan London, Rob Mackin, Nancy Plankey-Videla, Andrew Shrank, Spencer Wood, and Steven Wolf. In Philadelphia, Susan Lindee welcomed me to the history of science community at the University of Pennsylvania, where I found her enthusiasm for research wonderfully contagious. Among others at Penn, Riki Kuklick and Doug Massey gave me teaching jobs, Rob Kohler gave me sage advice, and Audra Wolfe—who later became my editor—invited me to parties. Since my arrival at Tulane University, the Department of Sociology staff, faculty, and graduate students have provided a collegial and encouraging environment for refining my arguments and putting the finishing touches on the manuscript. Special thanks to Donna Listz for administrative guidance, Sarah Kaufman for excellent research assistance, Barbie Stroope for eleventh-hour photocopying, and Melissa Abelev for unwavering help with the index.

Several people read some or all of this book or related material and provided valuable feedback. These include two anonymous reviewers, April Brayfield, Tim Brezina, David DeMarini, Tom Gieryn, Kevin Gotham, Neil Gross, Chris Henke, Sarah Kaufman, Daniel Kleinman, Kelly Moore, Maren Klawiter, Susan Lindee, Pam Oliver, Marline Otte, and Sara Shostack, who also directed me to key documents for the concluding chapter. My editor Audra Wolfe made fantastic suggestions toward first widening and then refocusing the book's contextual scope. I would also like to acknowledge the kind assistance of others at Rutgers University Press: Adi Hovav and Marilyn Campbell, for smoothing the transition from manuscript to book, and Robert Burchfield for excellent copyediting.

I met many scientists in the course of my research whose insight and encouragement have been invaluable. Those who generously submitted to my requests for interviews are listed in an appendix, but a few deserve particular mention. I am indebted to the generous kindness of Liz and Ed Von Halle, who insisted I stay with them during my three visits to Oak Ridge, Tennessee. Liz opened for me genetic toxicology doors I did not know existed and shared many a story from the glory days of the Oak Ridge Biology Division. I thank her for these and for the brown bag lunches. Fred and Christine de Serres, Herman and Marlene Brockman, and Phil and Zlata Hartman also invited me into their homes to share meals, memories, and evidence from their personal files. Others who trusted me with their stories are David Brusick, Jim Gentile, Michael Plewa, Elizabeth Wagner, John Wassom, and Errol Zeiger. At the offices of the Environmental Mutagen Information Center, John Wassom, Beth Owen, and the late Wilma Barnard were kind enough to provide me with a desk, access to the copy machine, and answers to all sorts of questions. At the Hoskins Library Special Collections at the University of Tennessee–Knoxville, Nick Wyman graciously helped me find what I needed. David DeMarini provided unflagging support for this project at every step. He not only gave me the longest interview of anyone I spoke with, but David also read and provided written comments on my entire dissertation, invited me to share some of my findings at an Environmental Mutagen Society conference in 2000, and provided valuable comment on parts of the book's conclusion. He also put me in touch with Andrew Kligerman, a geneticist at the Environmental Protection Agency's National Health and Environmental Effects Laboratory, who kindly created the images of chemical-induced chromosomal aberrations that grace the book's cover.

Financial support for this study was provided through a National Science Foundation Dissertation Research Grant SBR–9710776, a summer fellowship from the Tulane University Graduate School Committee on Research, and a research fellowship from the Georges Lurcy Charitable and Educational Trust. A much abbreviated version of Chapter 6 was published in 2001 as "The Environmental Mutagen Society and the Emergence of Genetic Toxicology: A Sociological Perspective," *Mutation Research* 488:1–8.

If the meat of this book lies in archives and interviews and its brains lie in the history and sociology of science, this book's heart springs from my family's love and kindness. For these I thank Carol Sue and Don Frickel; Shelby and Rocco Verretta; Anne, Dave, Devin, and Justin Divecha; Susan Lauffer; and Jim, Rebeckah, and Hannah Fussell. Mostly, though, I thank Beth Fussell. For her patience, strength, critical eye, and humor I am deeply grateful, and I dedicate this effort to her.

Chemical Consequences

1

Situating Genetic Toxicology

[A] human being contains hundreds of thousands of genes and although mutation in any particular one is exceedingly rare, the vast number of genes within an individual ensures that most of us carry one or two new mutant genes and that we are all, in fact, mutants.

−Testimony of Gary Flamm, NIEHS research scientist, at a U.S. Senate hearing on "Chemicals and the Future of Man," April 1971

In 1941 University of Edinburgh geneticist Charlotte Auerbach and her pharmacologist colleague John Michael Robson discovered chemical mutagenesis (Auerbach and Robson 1944). Their findings, based in wartime mustard gas research, provided strong evidence that highly toxic chemicals were capable of changing the genetic structure of living organisms. Over the next two decades, scientists enrolled a diverse array of chemical compounds—some naturally occurring, but most synthetic and highly toxic—as research tools in experiments designed to explore gene structure and function. By the time Auerbach published the first monograph on mutation research in 1962, chemical mutagens had become standard tools of the trade, occupying shelf space in research and teaching laboratories in universities, medical schools, and government agencies (Auerbach 1962b).

This disciplinary perspective began to yield to competing understandings when, in the late 1960s, a small group of geneticists began voicing fears that mutagenic chemical agents might pose serious, and possibly global, environmental threats. Outside of the lab, they argued, synthetic chemicals circulating among human populations represented genetic hazards that remained undetected in standard toxicological screens. As a result, potentially millions of people around the world faced daily genetic assault from exposure to what some scientists had begun calling "environmental mutagens." These were presumably safe chemicals used to grow and preserve food, produce cosmetics, develop pharmaceuticals, and manufacture countless other industrial products. Damage to an individual's sex cells initiated by environmental mutagens could, if passed from parent to offspring, remain within the population for generations and ultimately compromise the long-term integrity of the human gene pool. While Gary Flamm's remark in the epigraph that "we are all, in fact, mutants" was an important one, it was his concern over the generational and evolutionary consequences of chronic exposure to environmental mutagens that inspired his testimony before Congress. He was not alone. For the better part of a decade, these concerns mobilized scientists working in more than two dozen different fields to sustain a collective, multipronged campaign to avert what another scientist called a "genetic emergency" (Crow 1968).

Collective efforts to address the problem began in earnest with the creation of the Environmental Mutagen Society (EMS) in the early weeks of 1969.[1] Almost immediately, published research on chemical mutagens skyrocketed.[2] In June 1969 membership in the EMS was pegged at a modest 87; one year later the EMS claimed 452 dues-paying members—a more than five-fold increase. By the end of 1970, the society had published three issues of the *EMS Newsletter* providing a forum for scientific communication and debate, held its first scientific meeting that attracted 268 people, helped to establish a registry of chemical mutagens under the auspices of the Environmental Mutagen Information Center (EMIC), and sponsored several academic symposia and roundtable discussions on the genetic hazards of environmental chemicals (Wassom 1989). By 1976, scientists in North America, Europe, and Asia had helped establish eight additional EMS societies and four new journals. They also institutionalized annual conferences, training workshops, and colloquia for researchers and graduate students; they set up collaborative interlaboratory and interagency research programs, review panels, and funding mechanisms; and they developed dozens (eventually hundreds) of chemical mutagenicity bioassays that transformed the political economy of laboratory practice in mutation research.

Led by geneticists, these innovations were also intensely interdisciplinary, reflecting the efforts of scientists working in academic, government, and industry settings whose training was rooted in more than thirty disciplines and departments ranging across the biological, agricultural, environmental, and health sciences.[3] Although falling short of some scientists' personal visions of what this new science could become, their campaign had lasting impacts. Chief among these outcomes have been the emergence of a set of institutions, professional roles, and laboratory practices known collectively as "genetic toxicology." Today, much of what we know and don't know about the long-term genetic consequences of synthetic chemicals can be traced to scientists' collective attempts to transform chemical mutagens into an environmental problem.

This book is a historical sociological account of the rise of genetic toxicology and the scientists' social movement that created it. It covers the period 1910 to the mid–1970s but concentrates on events that occurred mainly in the United States between about 1968 and 1976. This narrower focus underscores the intensity and speed of the new field's consolidation. As late as 1966, neither genetic toxicology nor environmental mutagens had meaning in life sciences discourse. Although the environmental and health implications of pesticides were topics attracting increasing attention within the federal government, most geneticists—even those working in government laboratories—remained staunchly committed to basic research.[4] Ten years later Congress passed the Toxic Substances Control Act of 1976, legislation that included mutagenicity in EPA testing requirements for newly manufactured chemicals and that effectively created a formal market for genetic toxicology knowledge (Andrews 1999). By then, genetic toxicology had become an established field of practice, and environmental mutagenesis had become that field's dominant organizing principle. That transformation is my main subject.

Disciplines and Interdisciplines

A newly emergent environmental health science, genetic toxicology will not be familiar to most readers.[5] It emerged out of the mid–1970s as a science dealing primarily with "environmental mutagenesis," the study of how and why exogenous agents (mutagens) induce genetic and chromosomal changes (Wassom 1989). While mutagens can be chemical, physical (e.g., radiation), or biological (e.g., viruses), research in genetic toxicology has focused mainly on chemical-induced genetic change. In addition to laboratory bench work and field studies, genetic toxicologists also have been involved in developing policy-relevant approaches to genetic hazard identification, risk assessment, and pollution prevention (Preston and Hoffman 2001). These "basic" and "applied" modes for addressing the problem of environmental mutagens and mutagenicity historically have developed in tandem and to a large extent remain tightly if not inextricably intertwined.

Organizationally, the field is buoyed by a kaleidoscope of disciplines and research specialties. While most genetic toxicologists are trained in the molecular life sciences, some also have backgrounds in the health, agricultural, and environmental sciences (Hughes 1999).[6] The institutional settings of genetic toxicology are similarly diverse, spanning academic, governmental, industrial, and private nonprofit employment sectors.[7] Despite this diversity, genetic toxicology exhibits a distinct organizational structure. The institutional core of genetic toxicology lies in the federal science system of national laboratories and federal agencies. In the 1960s the Oak Ridge National Laboratory Biology Division in Oak Ridge, Tennessee, was the symbolic and institutional center of research that would become known as genetic toxicology. Since the 1970s, federal department and agency laboratories at the National Institute of Environmental Health Sciences, the National Cancer Institute, the Food and Drug Administration, the Environmental Protection Agency, and the National Center for Toxicological Research, among others, have played major roles in shaping the direction of research and testing in genetic toxicology.

As in many other areas of biology, form and content conjoin. Anchored in the federal science system, the linkages among academic, government, and industry scientists, as well as the symbolic and material lines connecting fundamental research on gene mutation to the development of bioassays, risk assessment studies, and chemical regulation policy, have become institutionalized in ways that constrain as well as enable the work that genetic toxicologists do, where they do it, and how that work informs environmental policy. Over the past thirty years, scholars have developed two main approaches for examining the problem of discipline and specialty formation. One is concerned mainly with institutions and the other mainly with discourses.

Institutionalist accounts of discipline and specialty formation in science identify the various social factors that shape the emergence of new knowledge fields (Ben-David and Collins 1966; Cambrosio and Keating 1983; Chubin 1976; Cole and Zuckerman 1975; Edge and Mulkay 1976; Griffith and Mullins 1972; Johnston and Robbins 1977; Kohler 1982; Krohn and Schafer 1976; Law 1976; Lemaine et al. 1976; Mullins 1976; Rosenberg 1979; Thackray and Merton 1972; Wallace 1989; Whitley 1974; Whitley 1976). Inspired by Joseph Ben-David and Randall Collins's (1966) comparative analysis of the development of psychology in Germany and the United States and by Pierre Bourdieu's (1975) structural theory of the "scientific field," this literature describes discipline formation as analytically distinct from, but motivated largely by, intellectual advances in scientific theory and methodology.[8] According to this perspective, disciplines are, in the main, institutional achievements. Whether or not discipline builders succeed and new disciplines become established categories in the intellectual division of academic labor depends upon some particular complex of social factors that shape scientists' strategic interests or "disciplinary stake" in the consolidation of new knowledge fields (Cambrosio and Keating 1983).[9]

In contrast, cultural studies of "disciplinarity" make few epistemological distinctions between the cognitive core of scientific knowledge and the social structures, practices, and processes that advance and suspend it (Klein 1996; Messer-Davidow et al., eds. 1993). Here science is examined in terms of highly fragmented local practices, shifting boundaries, and contingent meanings. Amid this heterogeneity, the problem of disciplinarity becomes one of finding "the means by which ensembles of diverse parts are brought into particular knowledge relations with each other" (Messer-Davidow et al. 1993:3). Cultural studies of disciplinarity invariably invoke the pioneering work of French cultural theorist Michel Foucault. Science is centrally implicated in the construction and maintenance of what Foucault (1980:131) called "regimes of truth"—the discursive, technical, procedural, and social-hierarchical apparatuses that determine which statements are accepted by society and made to function as true. "Discipline," from this perspective, refers to the coordination and exercise of power embedded within discourses and practices, often aimed at "the subjugation of bodies and the control of populations" (Foucault 1978:140).[10]

Recent books by Adele Clarke (1998) and Timothy Lenoir (1997) represent instructive efforts at bridging these competing frameworks and leave little doubt that the study of disciplinary emergence warrants renewed theoretical attention. Perched between disciplines, but also between theories of discipline formation, genetic toxicology is a good candidate for careful scrutiny. Like Clarke and Lenoir, I draw from both modes of analysis, but I am less concerned with what postmodernist science studies call the micropolitics of meaning than I am with the institutional politics of knowledge. This perspective views discipline building as a political process that involves alliance building, role definition, and resource allocation (Kohler 1982). The social construction of meaning is an important part of discipline building, but only a part. And while this study pays attention to social construction processes, it does so selectively. My main focus is on the structures and processes of decision making in science that influence who is authorized to make knowledge, what groups are given access to that knowledge, and how and where that knowledge is implemented (or not) (Guston 2000; Kleinman 2000). These are complex issues, made more complex by the fact that genetic toxicology is a knowledge field that maintains itself through regular and purposeful interaction with other fields and other domains: an interdiscipline.

Interdisciplines tend to exhibit considerably more organizational, economic, and epistemological variability than disciplines, and much of this difference flows from the historical contexts in which disciplines and interdisciplines have emerged.[11] In the United States, disciplines have tended to be anchored in university departments and maintain tight control of internal markets for the production and employment of Ph.D. students (Turner 2000). Interdisciplines are more likely to be located in less powerful (and thus less stable) institutes, centers, or programs and do not enjoy control of internalized markets.[12] Another important difference is that while disciplines tend to produce knowledge that deepens understanding of specific phenomena, which leads to increasing specialization, interdisciplinary knowledge is often guided by a collective interest in problem solving (Weingart and Stehr 2000:xii). These differences have political, institutional, and symbolic consequences. Born inbetween the laboratory and the world of social and political negotiation, genetic toxicology is a science nurtured as much by public culture as by professional practices, a knowledge form that thoroughly reflects the interplay of scientific research, institutional interests, and environmental knowledge politics.

Adherents to a cultural studies of science approach might insist that the flexibility, instability, and interactivity that characterize genetic toxicology do not distinguish it from more sharply bounded sciences. As Julie Klein (1996:4) observes, "In the latter half of the twentieth century . . . heterogeneity, hybridity, complexity and interdisciplinarity [have] become characterizing traits of knowledge"—which is to say that disciplinary boundaries are socially constructed, made to appear solid and stable when in fact they are riddled continuously by cross-border incursions of one sort or another.[13] Science simply fails to recognize this. Emily Martin (1998:26) argues that "what sets the sciences apart is that they claim to construct reality but not to be themselves constructed." Yes, the boundaries of disciplines and interdisciplines both are socially constructed, but they are constructed toward different ends. Interdisciplines are not simply infant disciplines in the process of "growing up." They are constructed to be—intended to be—something different. Interdisciplines do not feign the unity of purpose or theoretical cohesion that disciplines demand. To the contrary, the cultural boundaries that demarcate interdisciplines are intentionally ambiguous. Where the perception of stable professional boundaries and exclusive insider access to resources confers disciplinary status, interdisciplines exhibit boundaries that are purposefully porous and by necessity flexible. They are inclusive social forms, and that is the point.

Indeed, genetic toxicology's interstitial character and institutional ambiguity present what I consider a strategic analytical advantage. As Lynn Nyhart (1995:4) has observed, moderate institutional successes in science are "rarely written about, but may, in fact, be far more common than the more dramatic [achievements] of creating a new discipline." I also find that less is more. An analysis of genetic toxicology's modest successes can cast new light across the long middle ground of possibilities that lie between full-blown disciplinary success and utter failure.

How is knowledge produced, organized, and made credible "in-between" existing disciplines? What institutional conditions nurture interdisciplinary work? How are porous boundaries controlled? Genetic toxicology's advocates pondered similar questions. Some complained that disciplinary ethnocentrism prevented many biologists' appreciation for the broader ecological implications of their own investigations. Others worried about losing control of the "genetic thrust of the EMS" were the society to widen its scope to include "environmental effects other than the purely genetic."[14] Similar debate ensued over such issues as the efficacy of testing methods, appropriate standards of genetic risk, and the relative importance of distinctions between, for example, mutagenicity and carcinogenicity. Many people had a stake in how genetic toxicology came to be defined and controlled. I am interested in the specific ways in which these ambiguities and dependencies have worked themselves out; what those processes can tell us about the creation of interdisciplinary knowledge, practices, and careers; and how institution building in one small corner of biology was conditioned by larger economic, political, and cultural processes reshaping the nation.

American Biology on the Eve of Earth Day

The 1960s was a decade of massive transformation in American science, as in American culture, and in biology perhaps most of all. Science had emerged from the Second World War economically, politically, and culturally ascendant. The successful efforts of a conservative "scientific vanguard" (Kleinman 1995a:54) to secure federal support for basic science and to ensure scientists' research autonomy through the National Science Foundation, created in 1950, had sweeping implications for the way research was organized, funded, and practiced. Federal funding for basic research grew steadily, support from private foundations also rose, states expanded and strengthened university systems to accommodate rising numbers of college students, and scientists of varied stripes—but particularly physicists—gained new levels of access and authority within policy arenas (Geiger 1993; Kevles 1978).[15]

Philip Pauly (2000:239) has taken measure of the changes in biology marking this "long-term boom":

> The number of doctorates in all life sciences was virtually static from 1931 to 1948; by the late 1950s the number more than tripled, and then it tripled again in the next decade. The number of entries annually in *Biological Abstracts* quadrupled between 1940 and 1960. Between 1946 and 1964, NIH research grants grew from $780,000 to more than $529 million; federal funds for "basic research" in biological sciences (not including medical and agricultural research) swelled from $8 million in 1953 to $189 million in 1964.

This demographic and economic expansion fueled intense and widespread changes in the organization of biological and biomedical knowledge. A new generation of leaders like George Beadle and James Watson led a growing consensus that placed biochemical genetics at the center of the science of life (Pauly 2000). New money, new laboratories, and new technologies gave scientists ever greater access inside the genetic material and encouraged the development of new lines of basic research and new ties to the agricultural sciences and biomedicine. Emergent disciplines like molecular biology, immunology, and human genetics consolidated during this period, while older disciplines like genetics, cytology, and radiation biology found new footing to revitalize and expand.[16] As American biology entered an era of Big Science, American biologists became an increasingly visible presence in policy discussions and public debate on topics ranging from radiation fallout, cloning, space exploration, and biological weapons.[17]

Few places illustrate Big Biology during this period better than the Oak Ridge National Laboratory Biology Division—in many respects, the institutional birthplace of American genetic toxicology. Established in 1946, by the mid–1960s the Biology Division employed a massive scientific and technical workforce who pursued lines of research ranging from cytology and biochemical genetics to radiation protection and carcinogenesis (Johnson and Schaffer 1994). Under the guidance of its founding director, a cosmopolite biochemist named Alexander Hollaender who maintained close contacts with Washington's scientific and political elite, Oak Ridge became a center for world-class radiation biology that attracted visiting scientists from around the world.[18]

As the 1960s waned, however, the long economic boom in science finally, albeit temporarily, went south. Sentiment in federal government for unhindered basic research ebbed as interest in Congress shifted in favor of research that, like President Richard Nixon's "war on cancer," promised more direct social and health benefits and long-term economic growth (Dickson 1988:30).[19] Federal support for basic research fell 10 percent in real terms between 1968 and 1971 and then stagnated through 1976 (29). Thus a growing scientific labor force faced increased competition for available funding. At Oak Ridge and other Atomic Energy Commission (AEC) laboratories previously flush with federal funds, radiation biologists bemoaned the end of what Roger Geiger (1993:198) has called the "golden age of science."[20] For Hollaender and other "Oak Ridgers," budget constraints provided an opportunity and an imperative to retool their research programs, to forge new relationships with industry, and to look beyond "basic" science to research that promised nearer-term benefits. That Biology Division geneticists increasingly turned their attention to environmental health research is no real surprise given the political turbulence roiling beyond the shrouded hills and guarded valleys that sheltered, but did not completely isolate, Oak Ridge, Tennessee.

The politics of mass protest provided a different type of pressure that bore down on scientists during the 1960s and early 1970s. As Kelly Moore has shown (Moore forthcoming), the war in Vietnam politicized many scientists and graduate students, some radically so. Moore argues that among the many scientists who protested the role science played in that conflict, physicists and biologists predominated—physicists in part because of their discipline's economic connections to military-industrial research and their dependence on Department of Defense contracts, and biologists who understood better than most the human and ecological devastation that chemical defoliants wreaked on Vietnamese villages, farms, and countryside. But inasmuch as university campuses could be transformed into caldrons for mobilizing protest, campuses could also provide shelter from the storms of war. Among the first generation of scientists trained in environmental mutagenesis research were young men who found in graduate training in genetics an effective and respectable way to avoid the call-up.[21] The other politicizing force for these students, and not a few of their mentors, was a popular groundswell of environmental concern.[22]

Modern environmentalism exploded onto the U.S. political and cultural scene in 1970 with inaugural Earth Day gatherings that drew as many as twenty million people into the streets, onto campuses, and before houses of government across the nation (Mertig et al. 2002). In its wake, the ranks of existing national environmental organizations like the Sierra Club and the National Wildlife Federation swelled (Mitchell et al. 1992; Sale 1993), hundreds (eventually thousands) of new organizations came to life (Brulle 2000), industry went on the "counteroffensive" (Primack and von Hippel 1974), and politicians enacted a raft of federal legislation in response to the pending "environmental crisis."[23] This paroxysm of environmental action laid the cornerstone of today's "environmental state"—those laws, regulations, procedures, protocols, agencies, departments, and offices that structure ongoing efforts to manage natural resources, preserve wilderness, curb pollution, and protect the public health (Frickel and Davidson 2004).

Scholars have linked the rise of modern environmentalism to a broader cultural shift toward "postmaterialist" values that accompanied the rise of an increasingly mobile, educated, and consumerist middle class (Hayes 1987; Inglehart 1990). Yet as Christopher Sellers (1997:11) has observed, those "'values' would have remained empty without the new validity and concreteness that science brought to the threat of industrial chemicals." As he and others have shown (Gunter and Harris 1998; Palladino 1996; Potter 1997; Whorton 1975), concerted efforts among government and university scientists to better understand the environmental and health effects of synthetic chemicals anticipated the Earth Day mobilizations by decades.

Then as now, science occupied a paradoxical space within modern environmentalist discourse as nature's—and hence society's—enemy and savior.[24] The paradox is best illustrated by the 1962 publication of Rachel Carson's *Silent Spring* (1962), an event that many regard as a watershed moment (Brulle 1996). Not only did this slim book give forceful and radical expression to a growing body of disturbing scientific research, but Carson also in the same instant articulated what would become the dominant environmentalist critique of science to a mass audience. In the 1970s environmentalists took up Carson's critique, arguing that the institutions of science, either unwittingly or with complicity, aided and abetted ecological disruption. At the same time, environmentalists presented environmental problems in scientific terms as "real, important, pervasive" and also as "readily amenable to rational solutions" (Mertig et al. 2002:453).

Carson's contributions to the environmental movement, and the negative reactions they generated in the chemical industry and allied sciences (Graham 1970), are so well understood that they long ago attained the status of cliché. What is often not recognized, but which gains importance in the present study, is that in discussing "a long list of . . . chemicals known to alter the genetic material of plants and animals," Carson (1962:188) introduced the American public (and many scientists as well) to the idea of environmental mutagens. Moreover, because much of the existing research that Carson used to build her case was generated by geneticists and biomedical and public health specialists, genetic toxicologists a decade later would champion her work, often comparing it to their own efforts to draw public and government attention to genetic hazards. Finally, Carson's book helps to illustrate a major premise that infuses this study: that genetic toxicology bears the imprint (culturally, organizationally) of the field's often antagonistic and at times contradictory relationship to environmental organizations and activism. These antagonisms—real or imagined—did much to shape genetic toxicology's emergence as Earth Day dawned.

Origin Stories

Emerging sequentially from roughly similar (in some instances overlapping) milieus, the formation of human genetics—the application of genetics knowledge and practice to the study of human disease—provides an instructive comparative case for thinking about the developmental trajectories of new disciplines. Historians date its consolidation to around 1959 (Lindee 2004)—exactly one decade before genetic toxicology advocates founded the EMS. Like genetic toxicology, human genetics came together in a multidisciplinary fashion, geneticists and physicians approaching the genetics of human disease through different institutions, practices, and audiences. Also like genetic toxicology, the rise of human genetics was aided by charismatic leaders who established high-visibility programs in their home institutions and by technological developments in the neighboring disciplines of biochemistry and cytology.[25] Additionally, the emergence of both fields was marked by the creation of new knowledge markets and interaction with "external" social movements. Where the new politics of environmentalism gave a measure of legitimacy to genetic toxicologists lobbying for new rules regulating the use of mutagenic chemicals, so did an incipient abortion rights movement provide a legitimating context for human geneticists' promotion of prenatal screening and genetic counseling to expectant middle-class couples (Kevles 1985; Paul 1992, 1995). The key point of comparison for my purposes, however, is that both fields stabilized through a process of identity formation and institutional incorporation that was rapid and intense. Susan Lindee (forthcoming:1) describes this transformation in human genetics in terms of a cultural "criticality"—"the point at which interactive effects produce rapid institutional and social change." By 1960, "there were enough people and institutions around with a stake in genetic disease to support a series of related, and sometimes independent events, which have had profound consequences for the development of biomedicine."

Insiders' published accounts of the origins of genetic toxicology make a similar, if less explicit, claim, highlighting four main factors that in combination contributed to genetic toxicology's rise (Auerbach 1978; Brusick 1990; Crow 1989; Prival and Dellarco 1989; Sobels 1975; Wassom 1989). Predominant among them is a steady accumulation of facts and data through normal experimentation, punctuated from time to time by important discoveries that propelled research in new directions. Along the way, technological innovation such as chromosome staining techniques and the so-called Ames test made new methods of analysis possible as well as practical. Most accounts also acknowledge the key leadership role played by Alexander Hollaender in promoting genetic toxicology research and in laying the requisite organizational groundwork. He is broadly credited with possessing the visionary insight, administrative acumen, and selfless devotion without which genetic toxicology would not have happened when it did. The environmental movement provides the fourth causal factor advanced in insider accounts. It is during this period, common wisdom has it, that the countercultural critique of "technological society" and a related concern for the ecological and human health effects of environmental pollution reached a breaking point.

These accounts depict genetic toxicology as a science whose arrival circa 1970 was imminent, the result of distinct research streams shepherded toward a single point of confluence at precisely the moment that citizens' and government officials' awareness of the need for a new kind of environmental knowledge became most acute. This narrative gives the social origins of genetic toxicology a predominantly natural history in which only the timing of the streams' convergence seems to demand a cultural explanation.[26] But that cultural explanation also becomes naturalized, as if the drama were written and directed from somewhere offstage. Around 1969, the conventional accounts tell us, the transformation of the science of gene mutations into genetic toxicology was triggered by a vague external pressure called "environmental concern" (Crow 1989; Prival and Dellarco 1989).

This interpretation, if not wrong, is at least seriously incomplete. My analysis reveals very little to suggest that the rise of genetic toxicology was triggered by social conditions resembling institutional momentum or criticality that tipped the consolidation of human genetics a decade earlier. Even throughout the 1960s, despite thirty years of accumulating research on chemical mutagenesis, very little cross-laboratory comparative research took place. Research networks existed, but they were organized around experimental organisms (*Drosophila*, *tradescantia*, or *E. coli*), not around mutational processes. Geneticists brought toxic chemicals into their laboratories as research tools, not as environmental problems. The tipping point that insider accounts assume conditioned genetic toxicology's sudden consolidation did not exist. And thus genetic toxicology's emergence was neither inevitable nor coincidental.

How and why did genetic toxicology happen? I will argue that genetic toxicology was ushered in by a scientists' social movement. That is, collective action among scientists involved in genetic toxicology's creation was organized, strategic, and infused with environmental values; it elaborated a social critique of the disciplinary organization of science and sought to create a new way of ordering environmental knowledge. Building a science centered on the study of "environmental" mutagenesis and organizing a social movement of scientists committed to preventing further increases in the load of mutations were intertwined and mutually constitutive processes. During this period, for example, several new professional societies were created that formalized patterns of scientific communication and thus served scientists' professional research interests. Yet these societies functioned simultaneously as social movement organizations that helped to raise public awareness about the dangers of chemical mutagens and to advance problem-solving strategies to understand and minimize genetic hazards. In this respect, distinctions between science and public service blurred. At least for a time, activism and research were interdependent. The credibility of the movement depended on the credibility of the science and scientists backing it. Similarly, justification for the establishment of genetic toxicology as a new branch of genetics that served an explicit and direct public interest relied on convincing patrons, administrators, Congress, and (perhaps most important) biologists engaged in fundamental research that chemical mutagens were a problem that deserved immediate, systematic attention and continuous, long-term support.

The genetic toxicology movement's linkages to the larger environmental mobilizations at the center stage of American politics and culture during this period were complex and at times subtle enough to be easily missed. As I try to show, the scientists who established genetic toxicology were not merely responding to the environmental and public health concerns of the general public but were themselves active participants in those debates, often challenging the irrationality they believed was driving the contemporary environmentalist agenda. Even more, I will argue that genetic toxicology advocates developed a style of environmental activism designed to be effective where it would count the most—within the research contexts those scientists occupied on a daily basis. The symposia, technical workshops and training courses, review committees, and new professional organizations and journals, as well as the public lectures, congressional testimonies, and editorials that characterized the flurry of interdiscipline-building activity marking genetic toxicology's arrival, were in the aggregate a concerted, organized strategy to challenge the way research on chemical effects was conducted, by whom, and for what purpose.

Science and Social Movements

The idea that an intimate link exists between social movements and science is inimical to traditionalist students of scientific institutions such as Robert Merton and Joseph Ben-David who argued that science functioned as an essentially autonomous, self-regulating social system and that external interference in that system from states or other political actors was by definition corrupting (Merton 1973; Ben-David 1971, 1991). This proposition has been the target of sustained critique but has gained direct and indirect support in recent years from two unlikely sources: scientists who have assumed prominent activist roles in the "science wars" debates of the 1990s (Levitt and Gross 1994) and community activists engaged in social and environmental justice struggles (Bullard 1993).[27] The former argue that politics corrupts science; the latter maintain that scientists impede or co-opt grassroots struggles. Either way, but for different reasons, mixing science and politics amounts to a form of epistemological miscegenation that is best avoided.

For others, the "science and social movements" thesis appears far less heretical. It is familiar terrain for sociologists who study how movements and activists have encouraged reform, fragmentation, or retrenchment within professions or research enterprises by targeting the political character of expert knowledge (e.g., Amsterdamska 1987; Brown et al. forthcoming; Epstein 1996; Hess forthcoming; Hoffman 1989; Kennedy 1990; Nelkin 1977; Nowotny and Rose 1979). It is also familiar to historians of discipline formation whose work grants a similar shaping role to social movements. In Robert Kohler's (1982) history of biochemistry, for example, a movement to reform medical school education provided an important institutional opening for the teaching of graduate-level biochemistry courses in medical schools while, at the same time, biochemistry, a biomedical specialty similar in practice to the natural science disciplines, became an important symbol of intellectual rigor and purity that reformers used strategically to advance their agenda. The opposite dynamic is brought out in Susan Star's (1989) study of brain research in late-nineteenth-century Britain. In that case, a powerful antivivisection movement's opposition to live animal experimentation forced proponents of the localization theory of brain function to band together in defense of their experimental practices. In the process, the "localizationists" and their pet theory gained widespread attention and institutional credibility.[28] From this angle, social movements are important to science inasmuch as they present scientists and other professionals with opportunities to emphasize the social relevance of their research.

Other work views the importance of movements in the sciences and professions more metaphorically. Scholars writing in the sociology of ideas tradition use the language of intellectual movements and countermovements as heuristic devices to characterize moments of growing coordination and/or fragmentation within knowledge communities with respect to a particular set of research practices, claims, and identities (e.g., Camic 1983; Collins 1998). Rarely do these studies employ theoretical concepts from social movement theory to analyze these phenomena, however. Citing Nicholas Mullins's (1973) study of theory groups in American sociology, for example, Stephan Fuchs and Peggy Plass (1999:272) note that new science specialties often behave like social movements, "with tentative and unstable charismatic beginnings, slow and gradual institutional maturation, and, if the movement survives its liabilities of newness and adolescence, normalization into parts of the establishment." Examining the role of unconventional behavior as a force for institutionalizing change in formal bureaucratic organizations, Mayer Zald and Michael Berger (1978:824) argue for a "strong analogy" between social movements that occur in nation states and those that occur within organizations. Similarly, Rue Bucher and Anselm Strauss (1961:326) suggest that professions are comprised of "loose amalgamations of segments"—groupings that emerge within professions and take on different identities, interests, constituencies, and practices and that often emerge in opposition to existing segments. Like social movements, segments "tend to develop a brotherhood of colleagues, leadership, organizational forms and vehicles, and tactics for implementing their position" (332–333). In contrast to the social-shaping perspective, movementlike phenomena are theorized as normal features of institutional processes in science that "exist in even the most established professions and are the focal points of social change" (Bucher 1962:40).

Both approaches are relevant to my study. Like the social-shaping perspective, I am interested in how the politics of environmentalism influenced the rise of genetic toxicology. And like the movement-as-metaphor perspective, I also begin with the assumption that the dynamics of movements and the dynamics of new sciences can exhibit important similarities. Indeed, Fuchs and Plass's (1999) characterization of the instability and uncertainty that accompany new disciplines bears more than a passing resemblance to the early days of genetic toxicology as conveyed by the insiders' accounts summarized above. At the same time, however, normalizing movementlike processes in science can neutralize the political saliency of questions about when, where, and why typically conservative institutions of science undergo rapid and sometimes deceptively profound changes. I resist this tendency. Peering into this case through the lens of social movement theory, I examine how social movements in science are actually constituted.[29]

My analysis suggests that the transformation of mutation research into an environmental health science was more than a professionalization project or intellectual movement. The rise of genetic toxicology cannot be chalked up simply to business as usual. Although there can be little doubt that scientists' commitments to theories and the defense of professional identities and turf did influence this process, it is misleading to interpret collective action to create genetic toxicology as a "collective mobility project" motivated solely by disciplinary interests (Larson 1984:28). Group commitments extended beyond professional turf battles and involved changes in social and political values. Of course, genetic toxicology was shaped by the broader political and economic contexts in which those scientists lived and worked, but that, too, is just one part of a more complex story. The scientists who promoted a new way of thinking about chemicals and mutations were not simply responding to environmentalism, they were helping to create it. Similarly, political and economic pressures did not by themselves lift genetic toxicology into existence, nor was academic diplomacy sufficient to create and maintain the self-consciously interdisciplinary new science. It took something more. It took a scientists' social movement.

Political scientists and political sociologists reserve the term "social movement" for collective action that is organized around the expression of grievances and accompanied by demands for social or political change outside established institutional channels such as parties or trade unions.[30] It is the outsider status of social movements that scholars such as Sidney Tarrow find interesting. As he observes, "[t]he irreducible act that lies at the base of all social movements, protests, and revolutions is *contentious collective action*. . . . Collective action becomes contentious when it is used by people who lack regular access to institutions, who act in the name of new or unaccepted claims, and who behave in ways that fundamentally challenge others or authorities" (Tarrow 1998:3). Defining social movements as contentious politics immediately raises the question: if genetic toxicology really was set into place by a scientists' social movement and if social movements by definition involve disruptive politics, what was contentious or politicizing about the movement to create genetic toxicology? The scientists in my study did not lack regular access to the institutions of science. Indeed, many of the scientists that made up genetic toxicology's activist core held positions as editors, administrators, and laboratory directors who controlled access to funding and information. Many of their claims were based on established scientific knowledge, and their behavior was not atypical of discipline builders. They established new professional societies, organized technical workshops and symposia, delivered papers on the importance of genetic toxicology, and pursued research that advanced their institutional and theoretical interests. How was genetic toxicology contentious politics? At the outset, I can offer three related observations.

The first is simply that within the context of discipline-based knowledge production, advocating for interdisciplinary knowledge is a political act. At root the movement was a call for the development of a new field of applied biology at a time when biology was rapidly gaining legitimacy as a "hard" experimental science. The movement also advocated an interdisciplinary approach to mutagenesis and its attendant public health implications. The movement's goals thus were doubly contentious, urging geneticists to get involved in applied work and simultaneously to relinquish their exclusive claim to the problem of mutagenicity.

Second, the movement was contentious in that it involved a redistribution of disciplinary power. For Foucault, disciplining knowledge is primarily coercive, and the goal, ultimately, is social control. But as Christopher Sellers (1997:233) points out, discipline can also destabilize existing power structures. His research shows how, prior to 1960, industrial hygienists institutionalized a "scientific gaze" inside industrial workplaces, making it possible for researchers to examine workers' bodies for visible signs of occupational disease and hold industry and government more accountable for the workers' health (228). Similarly, genetic toxicologists have extended the scientific gaze even farther inside the body, to the level of the genetic material. They also have extended the scope of institutional concern beyond workers in factories to include consumers, communities, and the natural environment. They did so by experimenting on the genetic effects of "everyday" chemicals, developing new tools for estimating genetic risk, training toxicologists in genetics methods, establishing new professional societies, and engaging in public education and outreach. This conventional work had contentious implications. It gave new cultural and technical meaning to mutagenic agents, and it perforated institutional and ideological barriers that separated experimental work in genetics from public health and environmental politics.

In this sense, the story I tell about how a scientists' movement brought genetic toxicology to life bears important similarities to more recent efforts by science professionals to bring public attention to new varieties of chemical-induced ecological disruption, from ozone depletion and acid rain to "environmental hormones" and "environmental illness" (Hannigan 1995; Krimsky 2000; Kroll-Smith et al. 2000; Miller and Edwards 2001). Contained within genetic toxicology's history are lessons about the moral and political responsibility of contemporary knowledge communities as well as their capacity as historical agents to meaningfully address the long-term consequences of our industrial chemical culture.

Finally, in treating the topic of scientist activism, I am not so much concerned with why some scientists suddenly became motivated to attempt a reorganization of genetic knowledge and practice. I take the answer to that question as more or less self-evident: they became motivated by their understanding that the science of genetics could do more than it was doing to understand the causes, scope, and human impact of chemicals in the environment. This issue is interesting insofar as we assume that scientists are not or should not be political beings, not themselves interested citizens. But they are, and as history demonstrates time and again, they often act like it. Over the entire course of the past century, American scientists—as often as not biologists—have engaged in social and political action. Scientists have organized for socialism in American government and against Lysenkoism in Soviet genetics; they have promoted as well as opposed racist eugenics, as they have campaigned for academic freedom and against McCarthyism; they have worked to prevent wars and curb arms races and have lobbied against nuclear testing and for the Kyoto Protocol on Global Warming.[31] Some of the scientists appearing in this book were veterans of earlier movements for whom advocacy of genetic toxicology was a momentary detour away from different causes. For others, participation in the movement to establish genetic toxicology marked the beginning of activist-oriented careers or at least a politicization of their professional identities. Still others view the environmentalization of genetics research as a well-intended idea that mistakenly swerved too far afield of "real" science. These political biographies matter and are introduced in notes at various points throughout the book. More to the point, the varieties of activism undertaken by the scientists I studied should be understood as part of a longer historical tradition of political and social engagement by American scientists that continues today.

Overview, Sources, and Terms

How was the scientists' movement to establish genetic toxicology organized and sustained over time? My answer to that question structures the rest of the book. Chapters 2–4 focus on the political and economic contexts of mutation research and on the social networks and institutions that conditioned the mobilization of genetic toxicology scientist-activists. Chapter 2 describes how production logics, disciplinary values, and local institutional resources shaped early work on radiation and chemical mutagenesis. I examine the organisms, technologies, and techniques that geneticists and biochemists created or adopted to produce and study mutations in laboratory experiments. Despite its economic and theoretical potential, chemical mutagenesis failed to generate an autonomous disciplinary structure throughout the 1950s and 1960s. On the eve of genetic toxicology's emergence, very few organizations existed that could distinguish chemical mutagenesis institutionally from radiation genetics or, increasingly, from molecular biology. But things were changing. Chapter 3 considers the reasons behind an increase in support of environment-oriented research by the AEC, Congress, and science administrators. The institutional changes that signaled these shifts in research focus—what social movement researchers call "political opportunities"—set the stage for a period of intensive organizing action (1968–1973), when community identity among geneticists, biochemists, toxicologists, and others first began to coalesce into something different. Chapter 4 paints a general picture of the wave of scientist collective action during this period. It lays an empirical foundation for the next two chapters, which examine how scientist activism was organized and why it took the forms it did.

Justifying genetic toxicology meant convincing people that chemical mutagens constituted a credible environmental problem. This was not a straightforward task given that the genetic effects of everyday chemicals were anything but obvious. Chapter 5 examines how scientist-activists, faced with considerable uncertainty about the nature and scope of environmental chemical hazards, made the rhetorical case for genetic toxicology. In Chapter 6, my field of vision shifts from individuals acting collectively to organizations regulating collective action. I focus on the work and politics conducted through the EMS, the primary organizational actor promoting genetic toxicology in the United States. By specifying its functions as a boundary organization, first in demarcating a problem domain and then populating it with researchers from various fields, I am able to show how EMS officers and council members fused science and politics as a strategy for changing policy, pedagogy, and research practices.

Although the movement initially drew its moral, political, and professional legitimacy from the urgent need to prevent a "genetic emergency," the intensity of geneticists' concern over the long-term implications of environmental mutagenesis was not long-lived. By the 1980s, preventing heritable disease and preserving the genetic integrity of "future generations" had given way to competing raison d'être: understanding the causes of cancer. Chemical mutagens gained lasting importance in biomedicine not as environmental problems in and of themselves but as triggering mechanisms in carcinogenesis. In the book's conclusion, Chapter 7, I use this shift in the trajectory of genetic toxicology research as a backdrop for rethinking scientist collective action as a strategy for transforming environmental politics.

Data for this project derive from various sources. I have consulted numerous published documents, including scientific journal articles, editorials, "insider" histories, newsletters, conference proceedings, government reports, transcripts from congressional hearings, trade magazines, newspapers, and scientific information databases. Documents from archive collections, administrative records, departmental or program files, and scientists' personal files have also been indispensable sources of primary data. In-depth interviews with the scientists themselves provide the other main source of information for this study.

Most of the people I interviewed are or were academic or government scientists involved in chemical effects research. A few others work in industry or for environmental organizations. Many of these scientists have played important institutional roles as officers in professional societies, journal editors, program chairs, or directors of government laboratories. The interviews lasted from one to five hours. All but a few were tape-recorded. I transcribed some and made notes from the rest. They have been invaluable for gaining insight into the sequence of key historical events, actors, and impacts and for developing new leads and new lines of questioning. In addition to these formal interviews, numerous conversations with research and administrative assistants, graduate students, and archivists along the way have furthered my understanding of the institutional history of genetic toxicology.

In the end, I decided to use interview material as primary evidence only rarely. Wherever possible, I have used material sources or interviews with other scientists to verify the accuracy of scientists' statements of fact. In instances where I've been unable to locate supporting evidence for important claims, I note that as well. To preserve scientists' confidentiality, I identify the interviewee only in a few cases, where not doing so would add confusion. In those cases, I have received the interviewees' explicit permission to cite them directly.

One challenge in developing the descriptive narrative that structures my account has been the relative fluidity of terms used by scientists to identify the various intersecting fields of research in which they work and with which they identify on a professional basis. "Mutation research," "radiation biology," "mutagenesis," "chemical" or "radiation mutagenesis," "environmental mutagenesis," and "genetic toxicology" all are terms that could be and have been used to describe the field of research with which I am generally concerned. They are terms that are at once synonymous and different, distinct but overlapping.

Part of the confusion stems from the fact that it is difficult for participants and observers alike to pin down a moving target. An interdisciplinary field-in-formation is one in which multiple competing labels are used to identify it at any given point in time. In some measure, the labels scientists used depended on the audience they were addressing. Thus in 1970 when geneticists spoke to pharmacologists or toxicologists, they often used the term "genetic toxicology." When geneticists spoke to federal science administrators, "environmental mutagenesis" tended to be more commonly used. And when they spoke with one another, they used the older language of "mutation research," "radiation biology," or "chemical mutagenesis." These distinctions in audience, however, were never clear-cut, nor did they remain even nominally consistent over time. By the mid–1980s, for example, all things "environmental" had taken on a distinctly less palatable political flavor, and use of the term "environmental mutagenesis" itself became a contested label and point of debate among some researchers (and remains so today).

Another aspect of the problem stems from the fact that scientists' post hoc assessments of genetic toxicology's social and intellectual history tend to be influenced by the distinctions they draw between basic and applied research. For example, "mutation research" is a label that some scientists I interviewed reserved expressly for experimental research on the fundamental processes of gene mutation and chromosome aberration; they used "genetic toxicology," in contrast, to signify a distinctly applied science devoted solely to regulatory testing and risk assessment. Other scientists, when asked, told me that they saw little practical reason to separate the two and suggested that the differences at best are semantic, and at worst they are ideological. To avoid some of the inevitable confusion, unless otherwise noted I draw the following distinctions.

I use "mutation research" to refer to a science whose general subject is the study of mutagenesis—the processes of gene mutation and repair. Mutagenesis experiments made use of two general classes of mutagenic agents: radiation and chemicals. Thus scientists refer to "radiation mutagenesis" or "chemical mutagenesis" to indicate what kind of mutagenic agents were employed in their experiments. Before 1968, all of these terms were unproblematic and for the purposes of this study can be read at face value. Mutation research was generally understood by scientists as research that involved a variety of experimental organisms and using genetic and biochemical methods to address many of the core problems in developmental biology. Chemical and radiation mutagenesis were means to that end. Mutation research collectively provided much of the raw material—knowledge, technologies, and practices—for genetic toxicology. After 1968, however, that clarity of meaning dissolves, and it becomes important to draw two further distinctions.

I use "genetic toxicology" to refer generally to that complex of scientific institutions, knowledge, and practices used to identify and understand the effects of exogenous physical, chemical, and biological agents on genetic material. It is a science that began to assume an organizational form around 1968. Its relationship to mutation research is complex. It is similar to mutation research in addressing fundamental biological questions but goes further in addressing the public health implications of the unintended consequences of human exposure to mutagenic agents. "Environmental mutagenesis" is a term that came in to use at roughly the same time as "genetic toxicology," around 1968, and many scientists at the time used the two terms more or less synonymously. I make an explicit distinction, however. Where "genetic toxicology" refers to a scientific field, I use "environmental mutagenesis" to refer to a set of research practices, analogous in its relationship to genetic toxicology as radiation and chemical mutagenesis are to mutation research. I use "environmental mutagenesis" to signify the politicization of experimental research on gene mutations. In practice, environmental mutagenesis is technically equivalent to chemical mutagenesis—both employ chemicals to understand the nature and rate of genetic change. The key difference lies in the values underlying the choice of chemicals used in the experiment. In chemical mutagenesis research, scientists' choice of mutagens was guided primarily by interest in theory building. In contrast, researchers practicing environmental mutagenesis chose mutagens that were known pollutants and/or potential carcinogens. In this sense, the call for research in environmental mutagenesis signaled a politicization of the study of gene mutation and indicated a shift in the social values guiding experimental design.

2

Working on Mutations

Mutation has been essential to Mendelism at every step, not only as the source of heritable variations—round peas or wrinkled, vermillion eyes or the wild type, rough plaque or smooth—but as a tool for understanding.

–Horace Freeland Judson, *The Eighth Day of Creation*

"The central problem of biology is the nature of mutation," H. J. Muller announced to students and faculty at the University of Texas in 1916 (quoted in Pauly 1987:179). As the epigraph for this chapter suggests, assessments of mutation since Muller's lecture have become no less unequivocal. Yet if the central importance that geneticists have accorded to mutational phenomena in the biological sciences has gone largely unchallenged throughout the twentieth century and now into the twenty-first, it may be in part because, as Susan Lindee (1992:231) observes, "mutation has been a remarkably plastic concept, interpreted differently depending on the problem being investigated, the organism of interest, or the consequences of the interpretation."[1] As a cultural form in biology, the idea of mutation is ironically self-exemplified by its own historical mutability.

Where geneticists have tended to narrate the history of mutation research as a chronological recitation of methods employed in an "ever-narrowing, essentially linear search" for the gene (Wallace and Falkinham 1997:1; see also Auerbach 1976:1–14), this chapter follows in the tradition of historical science studies research on scientific practice and the material cultures of laboratory life (Clarke and Fujimura 1992; Hacking 1992; Pickering 1993). It places work on mutations and the role of chemical substances in that work in social context. But rather than offer a close-knit descriptive account of entanglements that obtain among the various elements—human and nonhuman—that populate laboratory sciences (Latour 1993), this chapter emphasizes the way in which practical interests of different actors shaped use of chemical mutagens in different ways, toward different ends, in different places. I am interested in how the "broader context of an economy of practices" for a time impeded disciplinary coherence and community among mutation researchers (Lenoir 1997:51; see also Kleinman 2003).

The first part examines how and why chemical mutagens became standard features in laboratory experiments. Along with ionizing radiation, geneticists and biochemists began to adopt powerful chemical compounds to produce and study mutations in the 1940s. As research tools, chemical mutagens held many of the advantages of radiation in permitting the mass production of mutations, but the variety of available chemicals and the heterogeneity of their specific effects also gave scientists a greater level of control over the production process than they had previously. There were direct and indirect economic benefits to this. Relative to radiation, chemical mutagens were inexpensive, easily obtained, and required little in the way of laboratory equipment. Consequently, scientists working in the agricultural sciences and in clinical medicine found potential in the capacity of chemical mutagens to produce "useful" mutations for plant and animal breeders, for pest control, and for cancer therapies. Despite this potential, chemical mutagenesis failed to generate an autonomous disciplinary structure throughout the 1950s and 1960s.

To help understand why chemical mutagenesis failed as a discipline, the second part of this chapter moves outside of scientists' laboratories to survey the institutional terrain in which research on chemical mutagenesis was embedded. Here, three factors are important to my argument. First, the newly elucidated biochemical nature of DNA gave biochemists new reason to pay attention to chemical mutagens and gave chemical mutagenesis a toehold in a rapidly expanding field, albeit one that mutation geneticists did not control. Second, the key journal in the field sought to integrate mutation research but did not distinguish chemical mutagenesis from other approaches. Third, the dominant research centers promoting chemical mutagenesis during this period were places in which radiation genetics remained the central research problem. As a result, on the eve of genetic toxicology's emergence, very few organizations existed that distinguished chemical mutagenesis institutionally from radiation genetics or, increasingly, from molecular biology. Thus, even as chemical mutagenesis practices expanded, local institutional interests continued to shape decisions about which mutagens scientists used and what aspects of the mutation process they studied. Ultimately, the mutation research political economy constrained the emergence of an autonomous chemical mutagenesis during the 1950s and 1960s even as it enhanced the conditions for the emergence of genetic toxicology in the 1970s.

Early Mutation Work

The science of genetics developed historically along two "quasi-independent" lines of research (Wallace and Falkinham 1997:1). One was concerned with identifying the factors of Mendelian inheritance. It asked, where and what is the gene? The second was concerned with the processes of inheritance. It asked, what do genes do? The experimental approaches developed during the early decades of the twentieth century to answer these questions centered on the study of genetic variation (Carlson 1966). Visible morphological changes arising from genetic mutations such as altered pigmentation or body shape were central elements in this approach. "While a trait common to an entire population could not be analyzed, a visually discernable deviation could be" (Lindee 1994:169). From their study of mutations, geneticists inferred knowledge of a gene's location on the chromosome and its modes of action.[2] But there was a problem.

Geneticists knew that specific kinds of mutations arising spontaneously were exceedingly rare. And studying rare events in a way that was statistically meaningful required access to very large populations of experimental organisms. The relative infrequency of any particular type of naturally occurring mutation thus meant that mutations had to be mass produced. In developing their methods of mutation production, geneticists selected experimental organisms carrying traits amenable to laboratory conditions and to the technical demands of experimental practice. Such an animal was the humble fruit fly, *Drosophila melanogaster*.

Drosophila's morphology favored the lab. Endowed with especially large chromosomes contained in its salivary glands, *D. melanogaster* was, in a sense, built to be seen. Scientists could view gross chromosomal aberrations using standard laboratory microscopes (Allen 1975). Living fast and dying young, the species' population dynamics gave it an additional competitive advantage over other laboratory organisms of interest to geneticists. Females bred regularly and quickly, their ten-day gestations routinely producing from 400 to 1,000 offspring. *Drosophila* was also ubiquitous. Of the more than 900 species in the genus *Drosophila* known to zoologists at the time, *D. melanogaster* was one of only eight native to all the world's major ecological regions. The biologists who went looking for it rarely had to travel far. Feeding on rotting vegetation and fruit, *Drosophila* thrived in densely populated urban environments and, conveniently, tended to be most plentiful in the fall, precisely when professors and students of zoology most needed the manipulable fruit fly for laboratory demonstrations in developmental biology.[3] As easy as *Drosophila* was to find in the wild, it was even easier to please in the "second nature" of the laboratory. It thrived in cramped quarters on old bananas, a food source that was inexpensive, easily prepared, and available to laboratory technicians year-round. Its domestic habits and nutritional needs, in short, made *Drosophila* a guest well suited to life in the laboratory (Kohler 1994:20–22).

Neither *Drosophila*'s laboratory domestication nor its consequent status as geneticists' experimental organism-of-choice during the first half of the twentieth century was automatic. While the natural qualities of both *Drosophila* and university zoologists initially drew them together in what Kohler (1994:19) describes as a relationship of "symbiosis," the fame of both required further mutual construction. *Drosophila*'s usefulness to geneticists depended on the development of a laboratory production system with the capacity to produce spontaneous mutants in sufficient numbers to permit quantitative genetic analysis. In order to construct the "standard" fly, the fly production process had to be improved, and those improvements also had to be standardized and disseminated.

During the 1910s and 1920s, Thomas Hunt Morgan's "Drosophilists" at Columbia University, after considerable experimental and organizational effort, succeeded at ratcheting up the rate at which mutant and wild-type flies were bred and crossed. In time, their artificially accelerated crossing schedules achieved economies of scale in which "the production of new mutants and new genetic knowledge fed on itself." The dynamics of the resulting production system were such that "the more mutants turned up, the more crosses had to be done to work them up. The more crosses were done, the more mutants turned up. The process was autocatalytic, a chain reaction. *Drosophila* became, in effect, a biological breeder reactor, creating more material for new breeding experiments than was consumed in the process" (Kohler 1994:47).

Despite the tremendous level of planning and organization involved in getting the *Drosophila* "breeder reactors" at Columbia and elsewhere up and running, spotting mutations still remained an extremely time-consuming and laborious task requiring a skilled eye and more than a little luck. The results of the Drosophilists' considerable investments in laboratory infrastructure and human capital, moreover, remained limited. During the period 1910–1926, *Drosophila* geneticists, students, and technicians had identified a combined total of only about 200 mutant flies (Carlson 1981:145).

The discovery rate of new mutants took a sharp upward turn in 1927 when workers in a single *Drosophila* laboratory found more than one hundred mutants in just two months. This exponential increase reflected the incorporation of a new technique into mutation work and marked a qualitative change in the logic of production in experimental genetics. Rather than increasing the probability of identifying naturally occurring mutations through highly regimented breeding and crossing schedules that propelled Morgan's biological breeder reactor, the newly introduced approach increased the rate of mutation itself. It did so by bombarding captive flies with x rays in order to *force* mutations, rather than waiting for mutations to appear spontaneously.

Radiation Mutagenesis

The groundbreaking work that opened the field of radiation genetics appeared in two papers made public in the summer of 1927. H. J. Muller, then a professor of genetics at the University of Texas and a former member of Morgan's "fly room" at Columbia, authored the papers. The first article, published in *Science* in July, announced Muller's discovery of x-ray-induced mutagenesis. The article was short on evidence but long on claims, among them the observation that treatment of *Drosophila* with x rays "had caused a rise of about fifteen thousand per cent in the mutation rate over that in the untreated germ cells" (Muller 1927; reprinted in Muller 1962b:246). The technique for achieving this astounding productive capacity was simple enough, and Muller's second paper, presented at the Fifth International Congress of Genetics in Berlin later that summer, delivered the details (Muller 1962c). The series of experiments involved exposing thousands of adult male and female *Drosophila* to x rays for varying lengths of time of up to forty-eight minutes. The flies then were mated with untreated flies and the offspring of successive generations examined for a variety of mutational lesions.[4] Many were found.

Neither Muller's audience in Berlin nor his peers in the United States failed to grasp the broader significance of these findings. As Muller's biographer describes it, "The paper created a sensation. The press dispatched the news around the world. . . . Like the discoveries of Einstein and Rutherford, Muller's tampering with a fundamental aspect of nature provoked the public awe. When Muller returned to the United States he found, to his surprise, that he was famous" (Carlson 1981:150). It is not difficult to understand why. The demonstration that the genetic material could be intentionally altered bore profound social and moral implications. For biomedical practice, the ability to destroy cells in a controlled manner meant a possible treatment for cancer. For clinical radiologists, the capacity of x rays to also paradoxically trigger the formation of cancer cells implied to some that the use of radiation in clinical practice should be highly restrained. For the eugenics movement, induced mutagenesis meant hope for those interested in controlling human evolution.

Muller was well aware of all these implications, noting various potential repercussions of x-ray mutagenesis in many of his early scientific papers on the subject. Indeed, there is some evidence to suggest that Muller's longtime interest in eugenic concerns was a factor in his decision to take up the x-ray experiments (Pauly 1987:179; see also Paul 1987). Muller also used this knowledge to motivate his political work. In his popular book *Out of the Night*, Muller (1935) advocated for a socialist-inspired "positive" eugenics program built on a foundation of equal rights for women, abortion on demand, birth control, state-sponsored child care, and voluntary artificial insemination—"using the reproductive cells of outstanding individuals as a means of spreading such socially desirable traits as high intelligence, cooperative attitudes, and longevity in good health" (Carlson 1981:228). Muller was also an early and vociferous opponent of the clinical use of x rays as a birth control technology and in noncritical diagnostic and treatment procedures (e.g., photographing fetuses or treating warts) (336–338) and during the 1950s became an outspoken critic of nuclear testing (Muller 1955a).[5]

Beyond its social and political significance, as a new tool for genetic research x-ray mutagenesis also had several important practical implications that bore directly on the political economy of experimental genetics. The experiments produced a wide range of "true" mutations across the length of the chromosome (Muller 1962a:246). Some of these were visible mutations that appeared to Muller to be identical to many of the "spontaneous" mutations already familiar to geneticists. This correspondence suggested to Muller that his system produced essentially the same products as did Morgan's breeder reactor, only much more quickly. Far more numerous than these "visibles," however, were lethal mutations. "Lethals" appeared as dominant and recessive mutations carried on the X chromosome of male and female *Drosophila*. Dominant lethals prevented germ-cell development and could be readily identified by a statistically significant increase in sterility in treated males. Recessive lethals prevented the development of progeny. These could also be easily identified by examining the egg sacs of treated females and counting the number of dead fetuses to calculate statistically significant changes in sex ratio and/or birth rate. Although geneticists at the time knew that lethals existed, they occurred too infrequently under normal laboratory conditions to be of use as a quantitative measure of mutational change. In Muller's studies, the prevalence of lethals offered scientists unambiguous evidence and accurate measures of a variety of gene mutations.[6]

Moreover, mutations induced by x rays could be predicted and therefore controlled to a degree not practically possible with time-consuming crossing methods. Muller's experiments provided evidence that a linear relationship existed between dose and mutation rate and also that radiation treatments administered at different stages of development resulted in different types of mutations. These findings suggested to Muller that now mutations could be made "to order." In that respect, x rays offered the geneticist as well as "the practical breeder" a relatively simple, quick, and cost-effective method for inducing specific kinds of mutations of interest to geneticists and plant breeders (Muller 1962a:251). Insofar as the new techniques facilitated the production of theoretically "rich experimental material" (Kohler 1994:162), x-ray mutagenesis represented not only a quantitative advance but a qualitative one as well. The wide variety of genic and chromosomal changes brought about by radiation treatments bore directly "on the problems of the composition and behavior of chromosomes and genes," making possible "attacks on a number of genetic problems otherwise difficult of approach" (Muller 1962a:249, 248).

Within a year other geneticists had applied the technique to a range of laboratory organisms with similarly impressive results (Carlson 1981:151). With widespread confirmation of the efficacy of radiation mutagenesis there occurred "an explosion of interest in the genetic effects of radiation" (164). By 1933, "most of the general rules of radiation genetics . . . were worked out" (Crow and Abrahamson 1965:263), and over the next three decades geneticists produced a "huge" literature cataloging the effects of ionizing radiation on complex organisms (Drake 1970:160). These practical and theoretical incentives contributed directly to the rapid rise of radiation genetics and largely explain why radiation mutagenesis became the dominant mode of mutation production in experimental genetics. In approaching this interpretation, it remains to consider briefly the role played by these new tools in facilitating this transformation.

Radiation mutagenesis operates through a single mode of action—ionization. The energy released by x rays penetrates the cell and causes physical damage to the genetic material. The ionizing action of x rays is essentially invariant. For the practical purposes that geneticists used them, an x ray was an x ray was an x ray. The main axis of variation was duration of the treatment, which researchers could easily control. Because the physical properties of ionizing radiation in mutagenesis do not vary from one day to the next or from one laboratory to another, x rays provided geneticists a research tool that was "ready-made," its standardization all but complete.[7] Moreover, by 1927 x-ray machines were already widely in use, having been incorporated into clinical and diagnostic practice by physicians since the late 1890s (Carlson 1981). If a genetics laboratory could not afford to purchase its own x-ray equipment, scientists could often arrange to irradiate one's experimental organisms at a nearby hospital or medical school. *Drosophila*, of course, were easily transported. The flies for Muller's first experiments, for example, were placed in test tubes and taken to a local "roentgenologist" who provided the x-ray equipment and administered the radiation treatments (Muller 1962a:251).

Thus radiation mutagenesis required almost none of the coordinating work and infrastructural development that was necessary in the construction of Morgan's breeder reactor. The universal properties of ionizing radiation and the widespread availability of x-ray equipment meant that, quite literally, "every laboratory worker could now produce unprecedented numbers of new mutations in whatever organism he was studying" (Auerbach 1961:234). Because less time was spent coordinating technical protocols and interpretive rules for identifying and categorizing mutations, the technique allowed for the rapid decentralization of mutant production. Whereas Morgan's breeder reactor worked in part because he and his students were working in the same place at the same time, Muller's contribution of x-ray mutagenesis went some way toward dissolving these spatial and temporal constraints.

Chemical Mutagenesis

In contrast to the excitement generated in 1927 by Muller's discovery of radiation mutagenesis, chemical mutagenesis began with a whimper. In autumn 1940 Charlotte Auerbach and J. M. Robson began a series of experiments at the University of Edinburgh in which standardized strains of *Drosophila* were exposed to an aerosol spray containing sulphide and nitrogen mustard compounds—mustard gas.[8] Their study, funded through a contract with the

Chemical Defense Establishment of the British War Office, was part of a larger effort by scientists in the United Kingdom and United States to explore the pharmacological and clinical aspects of mustard gas injury (Beale 1993).[9] The first successful results appeared in April 1941. A year later Auerbach and Robson delivered their initial report of these experiments to the British Ministry of Supply. The report was considered confidential, however, and British censors delayed full disclosure of their findings until the war's end.[10] In the interim, but still a full three years after completion of their initial experiment, Auerbach and Robson managed to publish a half-page research note in *Nature* (Auerbach and Robson 1944). That brief communication is remarkable mainly as an artifact of the British wartime security apparatus and less as an announcement heralding an important scientific discovery.

Making only vague reference to several "potent synthetic substances" that "produced mutation-rates of the same order as those obtained with X-rays," Auerbach and Robson summarized data from two experiments using allyl *iso*-thiocyanate, a naturally occurring oil produced from mustard seeds. Their experiments with mustard oil produced "a definite though slight effect on the mutation-rate" in treated male *Drosophila* (Auerbach and Robson 1944). In 1946 a second, equally concise letter in the same journal identified mustard gas (dichloro-diethyl-sulphide) as the chemical substance used in their experiments and described briefly the kinds of mutations and chromosome rearrangements it had induced (Auerbach and Robson 1946). Detailed reports of these ongoing experiments appeared in 1947 (Auerbach et al. 1947; Auerbach and Robson 1947), and mustard gas mutagenesis finally received a formal public introduction in 1948 when Auerbach presented a paper, "Chemical Induction of Mutations," at the Eighth International Congress of Genetics in Stockholm.

National security concerns and war gas research notwithstanding, geneticists' primary interest in identifying chemical mutagens lay in their capacity as research tools to shed light on fundamental questions in biology. In this, geneticists' motivations in the 1940s and 1950s were in essence identical to those underlying Muller's experiments with ionizing radiation in the 1920s. Auerbach underscored the theoretical significance of chemical mutagenesis when she told her Stockholm audience:

> The work on chemical mutagens is just at the beginning of its path. In my opinion, this path may be expected to branch out into various directions providing new approaches to old problems, as well as approaches to new ones which are bound to arise in the course of the work. Foremost among the old unsolved problems is the nature of mutation and of the gene. A comparison of the mutagenic effects of physical and chemical agencies, of the connection between chemical structure and mutagenic-

ity, of the way in which chemical mutagens exercise their effect, and of the interactions between various mutagenic agencies can be expected to throw light on this fundamental question. (1949:143)

Studies at the end of the decade confirmed both that mustard gas compounds produced mutations in a variety of standard laboratory organisms and that several other synthetic chemical compounds induced mutations in *Drosophila* (Loveless 1966:xii).

The search for chemical mutagens had in fact begun much earlier. Through a selection process that was "mainly random," geneticists during the 1920s and 1930s attempted to induce mutations using such potent industrial compounds as iodine, ammonia, and various metal compounds and known carcinogens (Auerbach et al. 1947:243). Muller, for example, had run experiments on *Drosophila* using lead arsenate, magnesium chloride, and an industrial dye called "Janus green," among other substances, before turning to x rays (Carlson 1981:153). Although a few of these experiments generated ambiguous results, the vast majority were clearly negative; none were clearly positive (Auerbach et al. 1947). Thus the mustard gas research did not so much spark interest in chemical mutagenesis as revive it.

The introduction of chemical mutagens into mutation research had impacts both greater and lesser than Auerbach's comments in 1948 at the International Congress of Genetics anticipated. There is no question that their discovery suggested new approaches to many of the central theoretical puzzles then confronting geneticists. These included the questions of whether proteins or nucleic acids were the central factors in genic replication, what were the chemical reactions that result in mutations, and what was the role of naturally occurring mutagens on evolution (Auerbach 1949:143, 1963).

But theory was not some abstract motor that drove the new field of chemical mutagenesis. At best, we should look to the theories as providing a rough guide or general scheme for understanding the field's early development. To give theory our full attention, as Auerbach herself has done in several retrospective articles and chapters published in the 1960s and 1970s (Auerbach 1963, 1967, 1978), is to fail to understand the lukewarm reception that geneticists gave chemical mutagenesis and the uneven incorporation of chemical mutagens into standard genetics practices over the next three decades. Despite the theoretical importance attributed to chemical mutagenesis, interest in the topic was sporadic, and the field remained near the margins of genetic research even after the flurry of promotional articles that Auerbach and Robson published in the late 1940s. If we are to appreciate more fully why this was so, we must look additionally to the material contexts of mutation work and to the political economic pressures shaping genetics research more broadly. For geneticists, the attraction of chemical mutagens lay as much in their capacity to increase the production efficiency of experimental systems as in their potential for refining or refuting theories of evolution, mutation, or the gene.

Chemical Mutagens and the Material Contexts of Research

Scientists' decisions regarding which chemicals were selected for analysis and how those chemicals would be utilized in the research process reflected the interplay of disciplinary and local pressures. The incorporation of chemical mutagens into existing research programs and experimental systems was shaped in part by a disciplinary imperative to maximize "useful" mutations at the heart of genetic science. As research tools, chemical mutagens functioned as a technical means toward that essentially economic end—the rational and systematic production of genetic damage. Here mutagens represented a form of capital in the genetics political economy in which the logic of production shaped the practical demands of research (Latour and Woolgar 1986). Production efficiency depended on geneticists' ability to get positive results, more often than not by using organisms and techniques already at hand and with which laboratory scientists had accumulated considerable experience. Thus local configurations of skills, knowledge, and resources also shaped definitions of the scientific utility and the corresponding value that scientists attributed to chemical mutagens.

Disciplinary Values

As the term came to be used in genetics research in the 1950s and 1960s, "chemical mutagen" was a label appropriate in theory to any chemical agent that produced alterations in the gene material. In practice, however, scientists reserved the term for those compounds whose chemical characteristics meshed with existing techniques and experimental systems in ways that produced positive results. To be useful to geneticists, chemical mutagens had to act selectively on the genetic material without destroying the cell or the organism in the process.[11] Scientists interested in studying subcellular processes had little practical use for chemicals found to be toxic to laboratory organisms. The converse situation bore similar implications: chemicals eliminated from laboratory organisms before penetrating the cell, or chemicals whose insolubility made for difficult application to treatment populations, also did not long hold geneticists' interest (Carlson 1981:260). In short, the fit between the chemicals and the technologies and methods of application used to measure their genetic and chromosomal effects goes some way toward explaining which chemicals geneticists considered valuable resources and which they did not. The experimental systems used to find chemical mutagens, in particular, tended to be very finely tuned to specific mutational endpoints, and it is in their design that we can best see how disciplinary values and interests became embodied in the tools and techniques that geneticists used to identify chemical mutagens and to describe their genetic effects. The *Drosophila* mutant *ClB* is a case in point.

Muller found the mutant that he named *ClB* in 1920 and over a period of several years developed the stock and standardized a method for its experimental use (Carlson 1981:117–119). *ClB* mutants (females) carried a rearrangement on one X chromosome that prevented crossing over and that had in addition a recessive lethal effect (the "Cl" condition; the "B" referred to the "bar eyed" condition, a dominant visible mutation). "The object of this procedure was to prevent crossing over and to kill off all sons of the tested F1 [first generation] females which carried this nontreated Cl chromosome" (Muller 1962c:262). In the crossing scheme Muller devised, brothers in the second generation of crosses (the F2) are mated with the treatment population—sisters carrying the Cl condition. If a lethal mutation arises on a sex chromosome, they are "detected by the absence from the progeny of a whole class of flies" such that "the corresponding culture will consist entirely of females—a fact which is, of course, readily observed even by an untrained person." By the time Auerbach and Robson used it in their original mustard gas experiments, the *ClB* test had become "famous" as a method for producing sex-linked lethal mutations.[12] And no wonder. Its design characteristics effectively eliminated the large amounts of human error incurred in classic methods of identifying visible mutations "without which the quantitative analysis of genetic radiation effects would have been impossible" (Auerbach et al. 1947:244). The same method, as we have seen, proved instrumental in opening the field of chemical mutagenesis twenty years later by filtering out or ignoring all forms of genetic and chromosomal damage save one.

The *ClB* test sharply restricted geneticists' empirical focus to one type of (lethal) mutation on the X chromosome among female flies. This reductionist strategy for finding and isolating mutations was replicated in many other experimental systems in *Drosophila*, the primary model used for chemical mutagenesis research (Auerbach 1978:182), as well as in several other eukaryotic systems in the decades that followed.[13] As in the *ClB* test, the practical goal in many tests was to find one chemical mutagen and/or its structural analogs that afforded a theoretically interesting view of the mutation process in a particular laboratory organism at a particular stage of its development. For example, with their positive results with mustard gas, Auerbach and her colleagues began experimenting with other chemicals with similar chemical structures or that produced similar pharmacological effects, soon identifying a class of chemicals that became known as "alkylating agents" (Auerbach et al. 1947:246).[14] In the mid–1950s, to take another example, with identification of the chemical nature of the gene as deoxyribonucleic acid (DNA), mutagens chemically related to DNA became fashionable (Auerbach 1961:235–236). The general point is that not just any mutation would do. The goal of test development in chemical mutagenesis during this period was to produce designer mutations that provided specific answers to basic questions concerning gene structure and action. Before Watson and Crick's double helix came to light in 1953, the chemical nature of the gene provided the question of central focus (Auerbach 1978).

As such, these systems were not designed to identify the broad sweep of potentially mutagenic chemicals. Nor were geneticists generally interested in testing chemicals individually for mutagenicity. That was not the end goal. As Auerbach commented in retrospect, "I personally did not feel tempted to test large arrays of chemicals for mutagenic ability. I knew that my chemical knowledge was quite inadequate to the task of forming my own working hypotheses in this field. I should have to test the hypotheses of chemists, and where would be the fun of this? . . . My own interest concerned the *process* of mutation" (Auerbach 1978:183). The logic of mutation production dovetailed perfectly with the disciplinary goals of classic genetics.

For scientists promoting a "genetic toxicology" in the late 1960s and 1970s, however, this logic would present constraints as well as opportunities. The existence of numerous experimental systems engineered to produce specific types of designer mutations in regard to nuanced theoretical questions gave geneticists concerned with public health issues only partial answers to the new question that piqued their interest. Namely, what are the genetic hazards posed to human populations by environmental chemicals? As we'll see in Chapter 6, this question embodied a somewhat different set of values for organizing research on chemical mutagens and for developing new mutagenicity tests.

Local Resources

When designing experimental systems for chemical mutagenesis research, geneticists tended to take the road most traveled, piggybacking on models designed for radiation genetics whenever feasible. Many, and perhaps most, of the bioassays used to conduct research on chemical mutagenesis in the 1950s and 1960s had been developed to use with ionizing or ultraviolet radiation.[15] The simple substitution of chemicals for radiation allowed scientists to capitalize on their working knowledge of mutagenicity bioassays and the genetics of the organisms with which they were most familiar, thereby increasing production efficiency at the local level.[16]

Another important set of resources in the mutation researcher's toolbox were the mutagens themselves. Although geneticists like Auerbach initially were attracted to the study of a few highly potent chemical mutagens for their "radiomimetic" qualities (i.e., comparable increases in mutation rates), the great potential of chemical mutagens as research tools lay in the manner in which they *differed* from radiation. To begin, chemical mutagens held a numerical competitive advantage over the few forms of radiation available to geneticists. The differences in kinetic energy output that could be marshaled in comparative studies of x-, beta-, gamma-, and ultraviolet-radiation waves paled in comparison to the interesting differences that could potentially be generated among the thousands of industrial chemicals then in existence (Crow and Abrahamson 1965). Geneticists realized early on that radiation caused relatively indiscriminate genetic and chromosomal damage. As we have seen, one of radiation's major strengths as a research tool is its capacity to, in effect, work the same way everywhere. The situation with chemical mutagens was exactly the opposite; their potential as research tools derived from their structural diversity and from the range of different effects produced through biochemical (rather than physical) modes of action.

Thus chemicals bore a distinct advantage over radiation in their capacity potentially to reveal *more* about mutational processes. For example, some chemicals had effects far milder than ionizing radiation. Rather than causing distinct breaks or mutations, these chemicals produced instabilities in the chromosome and the gene, which researchers termed "pre-breaks" and "pre-mutations," respectively (Auerbach 1978:184–185). Geneticists found in other mutagens a "storage effect," in which mutations were carried over a generation, appearing in the progeny of treated organisms (184). "Mutagen specificity," the ability of still other chemical mutagens to induce very precise chromosomal and genetic responses, provoked considerable interest among some geneticists.

Mutagen specificity was enough, in the minds of some, to legitimate chemical mutagenesis as a central pillar of mutation research. In her address titled "Past Achievements and Future Tasks of Research in Chemical Mutagenesis," delivered at the 1963 International Congress of Genetics, Auerbach (1963:279) argued that mutagen specificity "offers possibilities for analysis that so far have been almost wholly neglected." Certain chemicals' specificity of action suggested a range of comparative approaches that focused on differences in biological and biochemical responses "between species, strains within a species, sexes within a strain, cells within an individual, genes within a cell, and sites within a gene" (278). Auerbach outlined the ways in which chemical mutagenesis could be used to inform each stage of the mutation process, arguing, as she often did, for the importance of maintaining a *biological* view of mutation and against the reduction of mutation to biochemical interactions with DNA (279).

Throughout the mid and latter stages of her career, Auerbach ceaselessly advocated for a genetics and a molecular biology informed by cytological processes. In the preface to one of her textbooks, for example, Auerbach stressed "the need to realize that mutagenesis is much more than a physicochemical reaction of environmental agents with DNA. It is a biological process and, like all other biological processes, it is deeply enmeshed in the structural and biochemical complexities of the cell" (1976:xxvii). Such statements were a common feature in her public addresses and written papers. As biologists turned toward rather than away from a molecular understanding of life, however, Auerbach's vision of chemical mutagenesis research failed to materialize.[17] In other research contexts, however, mutagen specificity was exploited, and the productive potential of chemical mutagens was integrated into different systems of production.

Economic Mutations

The scientific utility of chemical mutagens extended beyond mechanistic studies of the mutation process. Although less explicit about the economic potential of chemical mutagens than Muller had been about x rays, Auerbach and her coauthors noted in a 1947 paper that chemical mutagens "with particular affinities for individual genes . . . not only would be of high theoretical interest but would also open up the long-sought-for way to the production of directed mutations" (Auerbach et al. 1947:243). To be sure, the potential economic relevance of chemical mutagens was not lost on geneticists working in select agricultural and biomedical sciences. In those research contexts, scientists tended to select mutagens with an eye toward maximizing the production of economically beneficial mutations, each following a logic tailored to a particular set of research and production goals.

At medical schools, hospitals, and the National Cancer Institute, scientists turned to chemical mutagens in the treatment of malignant tumors. Most valuable in efforts to develop chemotherapies were chemical mutagens that behaved like radiation in their ability to disrupt cell reproduction, but without the often severe side effects of radiation therapy. Early research in radiation genetics had shown that chromosome breaks that occur when cells are undergoing division are lethal but that nondividing cells often can withstand or tolerate chromosome breaks. This difference in the timing of the insult provided the basis for cancer treatments using chemicals to destroy rapidly dividing cancer cells without fatally damaging the healthy (nondividing) cells surrounding tumors (Auerbach 1976:256–257). Scientists proposed two approaches to the problem. One could find ways to make malignant cells more sensitive to insult so that less severe insult might achieve more effective results. Conversely, one could find ways to make normal cells more resistant to insults so that the attack on tumors could be stepped up without increasing the negative impacts on neighboring cells.

Plant breeders at U.S. land-grant universities became interested in chemical mutagens because of their potential for inducing a wide range of crop varieties. Mutational methods were significantly less costly and more rapid than traditional breeding methods (National Academy of Sciences Agricultural Board 1960), and chemical mutagens also bore a significant advantage over ionizing radiations. Alkylating chemicals proved more efficient than ionizing radiation in their capacity to induce smaller "point mutations" without also inducing more severe and indiscriminate damage such as chromosome breaks (Nilan

1973). Mutagen specificity meant something different to plant breeders than it did to geneticists working on the mechanisms of mutation. For plant breeders, the central problem was "phenotypic specificity," where the outcome was a phenotypic characteristic such as disease resistance or straw stiffness (in the case of barley). Here, the ends outweighed the means in economic importance. More fundamental "locus specificity" and the origins of mutational outcomes were less important to breeders (Auerbach 1976:453–454).

Economic entomologists used chemical mutagens in their efforts to develop nonchemical pest-control methods. "Chemical compound[s] which, when administered to the insect, will deprive it of its ability to reproduce" showed considerable promise for minimizing the biological and economic costs associated with autocidal pest control (Borkovec 1962:1034; Smith 1971). In the United States, these "chemosterilants" became the focus of an intensive research effort throughout the 1960s and 1970s that was promoted and conducted primarily by scientists at the Entomological Research Division of the U.S. Department of Agriculture (USDA) Agricultural Research Services.[18] Chemosterilants were less costly and more widely accessible than radiation sources. More important, chemosterilants offered greater flexibility in capitalizing on biological opportunities and combating the biological constraints inherent in the genetic systems of insects. On the one hand, "[s]pecies of insects appear in numerous polytypic forms, biotypes, races and strains," and this "genetic plasticity" made them extremely amenable to artificial genetic manipulation (van den Bosch and Messenger 1973:147). On the other hand, insects are not genetic equals. While some, such as screw-worm flies, could withstand sterilizing gamma-ray treatments, many others could not. Among other limitations, irradiation proved in many cases to seriously compromise insect health and/or mating behavior. In contrast, the vast range of available chemical compounds represented to scientists any number of potential alternatives to gamma and x rays. At least in theory, chemosterilants could be tailored to a particular pest's genetic specifications in ways that ionizing radiations could not.[19]

Chemical mutagens held great potential in genetics, plant breeding, entomology, and cancer science. In each of these contexts, mutagens functioned simultaneously as tools, theory, and commodities. The finished products of mutation work, be they cancer treatments or hybrid crop varieties, embodied values that reflected the intertwining of consumer and scientific markets. Yet despite their potential and realized contributions to biology, biomedicine, and agricultural science, chemical mutagens remained at the margins of most genetics research. Nearly twenty years after the announcement of Auerbach and Robson's mustard gas experiments opened the field, British geneticist Anthony Loveless would find occasion to remark that, "in spite of the evident potential of such substances for the elucidation of biological and especially genetical processes, biologists and biochemists have in general evinced a somewhat restricted and

fluctuating interest in the alkylating agents during the intervening years"
(1966:v). While Loveless was referring to only one of several classes of known
chemical mutagens, his observation was true of chemical mutagens generally.
Relative to radiation genetics in particular, published research on chemical
mutagenesis registered only marginal aggregate yearly increases (Wassom 1973).
One reason for the very gradual rate of growth during the field's first three
decades has to do with the mismatch between the highly efficient production
of mutation work inside laboratories and the organizational inefficiencies *among*
laboratories.

The Social Structure of Chemical Mutagenesis

Throughout the 1941–1968 period, chemical mutagenesis operated more like a
cottage industry than a rising scientific discipline or specialty. A disciplinary
structure complete with a professional society, journals, annual conferences,
textbooks, and graduate courses had long since come to define the field of radiation biology, and both the biological effects of radiation and the utilization of
radiation in the biological sciences (e.g., as radioisotopes) attracted immense
support from the federal government and from international agencies.[20] No
comparable levels of organization or government patronage existed in chemical mutagenesis. It was not until the mid–1960s that organizers of genetics conferences began to offer panels on chemical mutagenesis (as opposed to mutagenesis in general). Professional societies and journals explicitly promoting the
study of chemical mutagens and their genetic effects did not appear until 1969
with the formation of the EMS and its *EMS Newsletter*. And it was not until the
early 1970s that the newly created National Institute of Environmental Health
Sciences began to provide large-scale funding for intra- and extramural
research on the effects of mutagenic chemicals (NIEHS 1972:236–238). Thus few
organizational mechanisms existed for nurturing a professional identity among
chemical mutagenesis researchers that was distinct from radiation biology.
Absent the disciplinary structures that might have reasonably facilitated the
coordination of research programs beyond individual laboratories, the systematic study of chemical mutagenesis received little organizational attention.

But if discipline building in the conventional sense of forming a corporate
identity and organizational structure was on hold during this period, it was not
because the institutions and disciplines within which chemical mutagenesis
practices developed were moribund. Far from it. The "molecular revolution" in
biology, spurred by Hershey and Chase's demonstration that DNA was the car-
rier of genetic information (published in 1952) and the Watson-Crick model of
the double helix structure of DNA (published in 1953), effectively solved the
decades-old problem of mutation by revealing the chemical nature of mutagenesis (Auerbach 1976:2, 1978:186). In terms of kind, rate, scale, and complexity,

the transformations constitutive of the rise of molecular biology have been profound, and chemical mutagenesis was in every way caught up in social, cultural, and technical metamorphoses characterizing this "ultra discipline" (Abir-Am 1985:73).[21]

Through the mid–1950s, mutation work in genetics followed three general streams of research practice. As we have seen, two of these streams utilized ionizing radiation or chemical mutagens for studies conducted in higher plant and animal systems, primarily *Drosophila*. They were closely and importantly intertwined. The third research stream involved the use of ultraviolet radiation in studies of microorganisms, primarily *E. coli* bacteria and the bacteriophage viruses (Drake 1970:161; see also Creager 2002). Microorganisms like the phage T4 were advantageous as research organisms for several reasons, including a very simple and stable structure, genetic homogeneity, and rapid rate of growth and reproduction that made the examination of rare events like mutations even more accessible than in higher-order organisms (Drake 1970:5; Wallace and Falkinham 1997:20–22). Although there were exceptions, genetic work with microorganisms remained more or less distinct from chemical mutagenesis research, since each was organized around different kinds of mutation-inducing agents in different orders of experimental systems (Auerbach 1978:182).[22] With the elucidation of the chemical composition and structure of DNA, the material interests that defined chemical mutagenesis expanded and shifted their center of gravity, and these formerly separate streams of research practice began to merge.

Where the utility of chemical mutagens had once been understood largely in relation to ionizing radiation, chemical mutagens became increasingly important in terms of their structural relationship to proteins and nucleic acids (DNA). When the search for chemical mutagens reversed course, "the known nature of the gene [became] used for suggesting which substances might be mutagenic" (Auerbach 1963:276). Where researchers had once looked to chemical analogues of radiation or the highly potent classes of known chemical carcinogens to use as potentially revealing mutagens, researchers could now train their attention on analogues of DNA molecules and other chemical structures known to influence DNA. "It is on this [molecular] level," a hopeful Auerbach told an International Congress of Genetics audience in 1963, "that chemical mutagens are likely to remain useful tools for gene analysis. Their usefulness for the analysis of the mutation process has hardly been realized, but there are indications that it will be considerable. Very little has been done along these lines and many problems await solution" (1963:283).

Indeed, it was during this period of rapid change in the way that life itself came to be perceived and studied, and in the way life science came to be practiced and organized, that chemical mutagenesis found a firm toehold—not as an autonomous subfield of genetics but as a research niche in molecular biology.

By the time John Drake published his 1970 monograph, "The Molecular Basis of Mutation," scientists had identified a large number of chemicals found to be mutagenic in microorganisms. They constituted five classes of chemical mutagens, a class of mutagens that reduced the incidence of mutation (so-called antimutagens), and several "miscellaneous" mutagens that did not fit neatly into a chemical class (Drake 1970:ch. 13). Thus were chemical mutagens enlisted as foot soldiers in efforts to determine the molecular sequence of mutagenesis (Auerbach 1963:276–277). Chemical mutagenesis secured a niche in a rapidly expanding interdiscipline, but it was not, by and large, one of its own making.

The Organization of Communication

Mutation Research, a journal established in 1964 by a group of European and North American geneticists, quickly became the primary publication outlet for scientific studies of "mutagenesis, chromosome breakage, and related subjects" (as the journal's subtitle proclaimed). An international journal, it attracted contributions from North and South America, Europe, Asia, and Oceania. Although the center of gravity in mutation work remained in genetics, as genetic processes became increasingly relevant to adjacent fields and as the molecular study of mutation and nucleic acid biochemistry became more prevalent among geneticists, *Mutation Research* attracted an increasingly interdisciplinary readership.

The journal was established partly out of a shared concern that forces of specialization in mutation work were heading in the wrong direction—that is, away from a focus on genetic processes common to all living things and toward a focus on mutagenesis in specific organisms and on the development of specific experimental systems. The organization of genetic research at the Oak Ridge National Laboratory in 1955 illustrates the problem, with genetic research sections organized around experimental systems such as mouse, bacteria, and neurospora, rather than around the study of biological processes common to all or many of these organisms at genetic or cytogenetic levels (Table 2.1). Auerbach's (1962b) own monograph on mutation research methods is structured along similar lines, with eight of ten chapters devoted to methods in particular animal, plant, and microorganism systems. Some in the mutation research community, including Auerbach herself, came to perceive this situation as a potential hindrance to communication and understanding. As she later complained, "Mutation workers consider themselves specialists on bacteriophage, *E. Coli,* yeast, *Neurospora, Drosophila, Vicia faba* etc; not as being committed to the analysis of a special problem, even though their experiments may be restricted to one organism" (1976:xxvi).

The basis for Auerbach's concern was rooted in the double-edged facts of mutagen specificity. Experiments dating from the 1940s demonstrated that mutation is not one thing, or even one class of things, but instead a complex of

TABLE 2.1

**Three Genetic Research Sections, Biology Division,
Oak Ridge National Laboratory, 1955**

1. Cytology and Genetics

　　Cytogenetic effects of radiation

　　　　Paramecium

　　　　Tradescantia and *Vicia*

　　　　Maize

　　　　Neurospora

　　　　Timothy

　　Insect investigation

　　　　Grasshopper and *Habrobracon*

　　　　Drosophila Genetics

2. Mammalian Genetics

　　Genetic Effects of Radiation in Mice

　　Developmental Effects of Radiation in Mice

3. Radiation Protection & Recovery in Bacteria

Source: Oak Ridge National Laboratory Biology Division Files.

processes that have different and interacting mechanisms. In light of this compounded complexity, some scientists favored as a practical approach to the study of chemical mutagenesis one that would be explicitly comparative. While the specialization of research around a particular type or strain of organism could prove an advantage in more applied contexts—for example, in the development of genetic methods to control disease vector insects, where complete knowledge of the genetics, physiology, and behavioral ecology of a particular insect pest is essential (Wright and Pal 1967)—organismic specialization did not directly enhance the comparative analysis of mutation processes per se. Without comparative research, however, scientists were limited in their ability to infer genetic effects to higher organisms, including humans, based on mutagenicity studies conducted on lower-order organisms. This extrapolation problem, which has been a lasting point of contention among mutation researchers (Schull 1962) and is one that will receive focused attention in Chapter 6, was in part an outcome of the failure of mutation researchers to organize and pursue comparative work.

Mutation Research was an early attempt to counter this trend and to provide a center to an increasingly diverse and disparate area of research. In the preface to the journal's first issue, founding editor Frederik "Frits" Sobels, a geneticist at the State University of Leiden in the Netherlands, wrote of the need for a "unifying medium" to bring together researchers studying mutation in different organisms and publishing their work in different journals. Sobels saw much to be gained from an outlet for original research in which "mouse geneticists interested in repair phenomena may profit from developments in the microbial field, or people working on *Drosophila* may benefit from familiarity with problems of chromosome breakage in plants" (1964:1).

In the context of establishing a new journal that would appeal to an international and an increasingly interdisciplinary audience, *Mutation Research* provided space for many methods, experimental systems, and research tools (Table 2.2). Chemical mutagenesis was important to the editors of *Mutation Research* insofar as it represented, alongside radiation mutagenesis, chromosome aberration, and DNA repair, a set of common genetic processes. Similarly, chemical mutagens remained important as tools of laboratory production. But so did ionizing and ultraviolet radiations. If the contents of *Mutation Research* can be read as a microcosm of the type of research going on in the broader field, it is clear that even during the mid–1960s, chemical mutagenesis did not itself represent an axis of disciplinary organization or communication. Neither the preexisting orientation of mutation workers, who identified most closely with some particular laboratory organism, nor the *Mutation Research* editors, who sought to counter that trend by emphasizing function over form, favored the emergence of an institutionally autonomous chemical mutagenesis.

Institutional Dependency

In its development as a laboratory science, chemical mutagenesis was tightly intertwined with radiation genetics, involving many of the same people, methods, and laboratory organisms. With the obvious exception of the mutagens themselves, there were relatively few practical differences. The two social worlds were largely coterminous. For radiation geneticists, if not also for biochemists and other life scientists, taking up chemical mutagenesis involved few technical or financial costs.

Low barriers to entry into chemical mutagenesis research obtained in part because, regardless of the institutional context, the scientific work involved tended to be small scale. Experiments could be run by a few people working together in a single laboratory because material production requirements were easily met. Auerbach's (1947) early mustard gas experiments, for example, involved just a few hundred flies reared in screen-covered test tubes. The shift toward microorganisms further reduced the time and expense involved in chemical mutagenesis experiments.[23] This type of research also required relatively little investment in special equipment or materials. Unlike radiation genetics, chemical mutagenesis did not require the use of expensive x-ray

TABLE 2.2
**Research Tools Used in Experiments Published
in *Mutation Research*, vol. 1 (1964)**

Organism	No. of experiments	Mutagenic Agent	No. of experiments
Viruses		Ionizing radiation	
Bacteriophage	3	X rays	15
Actinophage	1	Beta rays	1
Bacteria		Gamma rays	4
E. coli	6	Ultraviolet radiation	7
Salmonella	1	Chemical mutagens	
Haemophilus influ.	1	Alkylating agents	11
Paramecium	1	DNA analogues	7
Fungi	1	Other	7
Yeast	2	Heat/kinetic energy	3
Plants		Gas	2
Vicia faba	5		
Other	4		
Insects			
Screw worm fly	1		
Drosophila	17		
Mammals			
Mouse	3		
Human	1		

machines, the space to house them, or the technicians to maintain them.[24] Chemical samples were often available from departments on campus or could be obtained free of charge from chemical companies or other scientists upon request. Authors' acknowledgments, for example, to "Chas. Pfizer and Co., Inc. for a gift of streptonigrin" (an antibiotic) (Kihlman 1964:61) or "to B. A. Kihlman for the gift of the 8-ethoxycaffeine" (Scott and Evans 1964:155), were common. Standardized organisms were either bred within one's own laboratory or obtained free from another lab or from one of the various stock centers. In short, geneticists could come and go in the world of chemical mutagenesis more or less as they pleased, confident that the transition from radiation work to chemical work (and back again) would generate relatively few expenses, financial or otherwise.

Although the number of chemical mutagens that scientists identified and studied continued to accumulate, throughout the 1950s and 1960s those increases were largely sporadic. Beyond the local level of the laboratory, chemical mutagenesis research was less systematic, with little coordination of experiments between laboratories. As a result, knowledge accumulated in piecemeal and often haphazard fashion. This situation prompted mutation geneticist Frederick J. de Serres (1981:1) some years later to remark that "in the past, data on chemical mutagens has been generated and published in the scientific literature on a more or less random basis. Individual chemicals enjoy a brief period of 'popularity' that leads to a burst of publications in the same or sometimes related assay systems." The relative lack of coordination or integration of interlaboratory research had important medium-term impacts. "The incompleteness of the data base," de Serres continued, "in many of these cases, makes comparative mutagenesis difficult or impossible." Low transition costs alone, it seems, were insufficient to ensure the steady increase of research interest in chemical mutagenesis.

Thus, while there may have been few practical or economic impediments to dissuade geneticists from pursuing research involving chemical mutagens, there also were few institutional incentives for organizing a systematic attack on the problem. What chemical mutagenesis possessed in terms of theoretical potential or economic utility for geneticists working in both basic and applied contexts, it lacked in terms of autonomous organizational resources. The field, as it existed, had no institutional center of its own. Instead, chemical mutagenesis remained indelibly intertwined with and to a large extent institutionally dependent on radiation biology.

Articles published in *Mutation Research* from 1964 to 1968 shed some additional, if indirect, light on this structural dependency. Despite its widely international appeal, research productivity in mutation work concentrated in just three countries (Table 2.3). Scientists from Great Britain, the Netherlands, and the United States authored nearly 60 percent of the articles published in *Mutation Research* during that period.[25] Within each of these three countries, article output concentrated heavily in three research institutions. Twenty percent of the articles originating in Great Britain were contributed by researchers at the Institute of Animal Genetics at the University of Edinburgh in Scotland. In the Netherlands, scientists from the Department of Radiation Genetics at the State University of Leiden contributed 39 percent. And in the United States, researchers in the Biology Division of the Oak Ridge National Laboratory accounted for 27 percent. The concentration of research productivity marks these three sites as centers of mutation work during the 1960s.

The laboratories at Edinburgh, Leiden, and Oak Ridge served as institutional anchors in Europe and North America for the field of mutation research.[26] Importantly, they also represented simultaneously the institutional

TABLE 2.3

Research Articles in *Mutation Research*, by Country (1964–1968 cumulative)

Country	Articles Credited*
United States	121
Great Britain	79
Netherlands	28
Italy	24
Sweden	21
W. Germany	20
Japan	15
India	12
Canada	9
France	9
Israel	9
USSR	8
Australia	8
Belgium	8
Other†	24

*Total articles = 385 The column total (n = 395) is greater than the actual article count because multiple countries received credit for international collaborations (n = 10).

†E. Germany (5), Czechoslovakia (4), Switzerland (4), Brazil (3), Finland (2), Argentina (1), Bulgaria (I), Denmark (1), New Zealand (1), Norway (1), Romania (1).

core of chemical mutagenesis and, as such, do well to illustrate the institutional intertwining of radiation and chemical mutagenesis as social forms of scientific practice. The historical development of each of these institutions involved an early focus on radiation genetics out of which grew ongoing research programs on the comparative effects of ionizing radiation and chemical mutagens. Additionally, each institution possessed strong ties to national research bodies that encouraged mission-oriented science in the national interest. Finally, the three radiation geneticists who directed administration and research at these laboratories—Charlotte Auerbach at Edinburgh, Frits Sobels at Leiden, and Alexander Hollaender at Oak Ridge—were longtime advocates of chemical mutagenesis at their home institutions and beyond.

Originally named the Animal Breeding and Genetics Research Organization, the Institute of Animal Genetics was an animal-breeding research center set up by the British Agricultural Research Council at the University of Edinburgh to improve agricultural output and to lessen Britain's dependence on food imports (Falconer 1993:137). Although Auerbach's work on mutagenesis began earlier, radiation genetics at the institute began formally in 1947 with the formation of a small research group working on the mutagenic effects of radiation in mice. In 1958 the working group was reconstituted as the Mutagenesis Research Unit, with Auerbach appointed as unit director (141). Under Auerbach, this unit pursued extensive comparative research on the mutagenic effects of ionizing radiation and chemicals (Kilbey 1995). As noted, Auerbach argued repeatedly throughout the 1960s and 1970s against relying solely on a biochemical view of mutagenesis. She saw chemical mutagenesis as a way to connect the biochemistry of nucleic acids to processes occurring at the genetic, cellular, and organismal levels.[27]

The Department of Radiation Genetics at the State University of Leiden was the first and only department so named in the Netherlands (Sankaranarayanan and Lohman 1993). In 1959 members of the department elected Frits Sobels chair, a position he served until his retirement in 1987. Sobels had studied with Auerbach at Edinburgh and had exchanged several rounds of letters with H. J. Muller, then at Indiana University. Both are said to have had profound influences on his thinking and institution-building efforts, which, as we have seen, included a major hand in the creation of *Mutation Research* in 1964 (Bridges 1993). Like Auerbach, the research program Sobels pursued for most of his early career centered on comparisons of radiation and chemical mutagenesis in *Drosophila*. By 1975, the newly renamed Department of Radiation Genetics and Chemical Mutagenesis reflected this dual emphasis.

Ties to mission-oriented research in the national interest were strongest and most apparent at Oak Ridge. The Biology Division at the Oak Ridge National Laboratory was a major component of the AEC's efforts following the Second World War to harness radiation in the national interest during peacetime. Under the strong administrative hand of Alexander Hollaender, the Biology Division grew into one of the premier biology research centers in the world. Unlike Auerbach and Sobels, both of whom were trained as *Drosophila* geneticists, Hollaender received graduate training in physical chemistry, and his most important scientific research involved studying the effects of light waves on a number of experimental systems, from viruses to higher plants and animals. Hollaender was among the first to understand that mutations were associated with changes in nucleic acid, at a time (1930s) when most geneticists thought that proteins were the critical genetic molecules. Thus he approached chemical mutagenesis with a deep sense of the importance of nucleic acids to the mutation process (Setlow 1987). The research division Hollaender built at Oak Ridge,

TABLE 2.4

**Research Articles in *Mutation Research* from U.S.,
by Institution (1964–1968)**

Federal research institutions	*49*
Atomic Energy Commission laboratories	
Oak Ridge National Laboratory	33
Brookhaven National Laboratory	4
Argonne National Laboratory	2
National Institutes of Health	
National Cancer Institute	1
National Institute for Neurological Diseases and Blindness	4
USDA Agricultural Research Service/ Experiment stations	3
Food and Drug Administration	1
Smithsonian Institution, Division of Radiation and Organisms	1
Academic research institutions	57
Medical and public health schools	16
Life sciences	37
Agricultural sciences	4
Private research institutions	15
Nonprofit laboratories	14
Industry laboratories	1

which centered on genetic research and nucleic acid biochemistry, reflected this insight. At its peak, the Biology Division employed 450 scientists, technicians, and administrative workers, making it the largest division at Oak Ridge and larger than all the biology divisions of the other national laboratories combined (Johnson and Schaffer 1994:115; Setlow 1987; von Borstel and Steinberg 1996:1052). It was a place rich in material resources, "unique" during the 1950s "for the variety of organisms used for genetics experiments," where only parking space seems to have been in short supply (von Borstel and Steinberg 1996:1052). Hollaender's Biology Division not only produced knowledge about mutagenesis; through its postdoctoral and visiting scientist programs that drew scientists from Western Europe, Canada, Asia, and Latin America, it also produced mutation researchers in vast numbers. Overflowing with human, financial, and material resources, Oak Ridge embodied many of the key forces of production in mutation research, effectively grounding the political economy of mutation research in the United States securely to the federal science system (Table 2.4).[28] Increasingly, genetic research at Oak Ridge emphasized chemical mutagens and mutagenesis. By 1965, research teams at Oak Ridge were involved in experiments on chemical mutagenesis in microorganisms, fungi, and paramecium.[29]

The key sites for mutation research also were the most important places for chemical mutagenesis. At Edinburgh, Leiden, and Oak Ridge, research programs in chemical mutagenesis grew directly out of radiation genetics. In these research centers, and in the efforts of the individuals most centrally identified with these institutions, few organizational distinctions between chemical mutagenesis and mutation research more generally are apparent. For the most part, money, people, laboratories, organisms, and mutagens intertwined in practice. As a result, there were very few organizational structures in place to distinguish chemical mutagenesis institutionally from radiation genetics or, increasingly throughout the 1960s, from molecular biology. At the same time, the advance of molecular biology and the increased emphasis on the molecular mechanisms of mutation that accompanied it brought increased interest to chemical mutagenesis as a research tool, but that interest was diffuse and highly decentralized. Thus, as chemical mutagenesis expanded in practice, its social coherence as a field in itself weakened.

Conclusion

In this chapter, I have described the development of mutation research during the 1940s, 1950s, and 1960s in order to understand the research contexts that conditioned the rise of genetic toxicology in the 1970s. The analysis has focused on work, tools, institutions, and the relationships (economic and otherwise) that connect them.

Much like the ionizing radiation that came before, the discovery of chemical mutagenesis gave scientists the ability to achieve economies of scale in the production of mutations not possible through earlier cross-breeding techniques. In addition, the wide variety of chemical mutagens then in existence provided even greater means of generating specific mutational types. This represented a particularly powerful way to control the mutation process, one that held potentially significant benefits for both basic and applied research. As important a research tool as chemical mutagens seemed to be, the field remained tethered to other problem domains, first in radiation genetics and later in biochemistry. But the reasons why an autonomous disciplinary structure did not emerge around chemical mutagenesis have less to do with the state of knowledge or theoretical significance of chemical mutagenesis than with the material nature of chemical mutagens themselves, on the one hand, and with organization of mutation research, on the other.

The enormous heterogeneity among chemical elements and compounds found to induce mutagens meant that mutations could be practically tailor-made around local research interests in particular organisms and specific mutational end points. This menu of choices, along with the small-scale economic requirements and low transition costs that geneticists faced in moving from radiation to work with chemicals, also meant, however, that there seldom were pressing reasons to engage in cross-laboratory collaborative or comparative work. As a result, the organization of experimental practice in mutation research came to resemble a cottage industry that reflected the knowledge interests of individual scientists and their local institutions as much as the more general concerns raised in genetics theory. Where chemical mutagenesis did connect to larger disciplinary structures, as in molecular biology, it did so as a means to an end rather than as an end in itself. Similarly, at those research centers like Hollaender's laboratory at Oak Ridge that did the most to promote chemical mutagenesis, the field remained institutionally dependent on research programs driven by radiation genetics. Where in the 1940s chemical mutagenesis was driven primarily by local political economies, by the mid–1960s the field was driven increasingly by extralocal political economies organized around DNA biochemistry and radiation genetics. Although the axes of fragmentation were changing, the field remained highly decentralized, with few institutional markers to distinguish chemical mutagenesis as an autonomous field of knowledge.

The institutional fragmentation of chemical mutagenesis also helps explain why concern about the human health effects of chemical pollution arose so late among geneticists—those people almost uniquely situated to understand both the magnitude and the depth of the problem. These institutional constraints and opportunities are the subjects of the next chapter.

3

Making Room for Environmental Mutagens

> It is tempting to consider the possibility that one of the means by which evolution adapts mutability to environmental requirements is the achievement of a balance between the production of mutagens and sensitivity to them.
>
> —Auerbach, Robson, and Carr, "The Chemical Production of Mutations"

This chapter's epigraph, appearing in the closing paragraph of one of Auerbach's early papers on mustard gas mutagenesis, comes deceptively close to connecting chemical mutagens existing in nature to changes in the processes of evolution. Had the paper made that link, the role of environmental mutagens in upsetting evolutionary fine-tuning might have gained an explicit purchase in the literature on chemical mutagenesis from the start. As it turned out, generalized scientific concern over the genetic implications of exogenous chemical mutagens lay dormant for nearly thirty years. This chapter seeks to explain the latent period between the discovery of chemical mutagenesis and its politicization three decades later. Why did the scientists' social movement that began its rapid and largely successful bid to constitute genetic toxicology around 1969 not mobilize sooner?

Three answers immediately present themselves, but none, I think, makes a strong sociological case. The first answer is that the requisite knowledge simply did not exist until late in the game, but once it did, geneticists quickly began to lobby for policy action. This convenient argument withers considerably in the face of historical evidence. New knowledge does not explain the rise of genetic toxicology. The fundamental claims that genetic toxicology scientist-activists made about chemical mutagenesis in 1969 were common knowledge in genetics much earlier. Geneticists knew that most mutations are recessive and so will remain in the gene pool for many generations (Muller 1927). They also knew that mutations are almost always deleterious, and many believed that an increase in the "mutational load" in a population will have negative long-term consequences for the genetic integrity of a species (Muller 1950).[1] Geneticists knew that mutation could involve biochemical processes, and this was borne out in research demonstrating that many different chemicals could induce mutations and chromosomal aberrations in laboratory organisms (Auerbach 1949). While there can be no doubt that new knowledge and new test systems produced throughout the 1960s strengthened the case for genetic toxicology, the empirical and theoretical basis of those claims predated its formation by at least fifteen years. Indeed, as we will see, some geneticists harbored these very concerns as early as 1950.

A second answer is that most geneticists involved in mutation research believed that only highly toxic chemicals posed a genetic threat and that human populations would not be exposed to these substances except through accident or an act of war (Auerbach 1978; Crow 1989; Wassom 1989). Here again, the historical record raises questions this thesis cannot easily answer. While it was the case that most research on chemical mutagenesis utilized potent chemical compounds such as the so-called alkylating agents, experimental work on nontoxic, mildly mutagenic substances such as caffeine existed in the published literature quite early on (e.g., Fries 1950). Additionally, other compounds such as the synthetic chemical pesticide DDT, which the analytical chemists and entomologists at the USDA knew in the late 1940s to be both highly toxic to humans and in widespread use (Gunter and Harris 1998), seem to have been generally ignored by mutation researchers. "It is apparent from the literature that there has been no large scale testing of pesticides for mutagenic activity" one report noted as late as 1969 (U.S. Department of Health, Education, and Welfare 1969:611).

The third suggestive but ultimately insufficient answer is that efforts to study the genetic effects of ionizing radiation, a program intensively encouraged and supported in the United States by the AEC and the National Research Council and by international agencies such as the World Health Organization (National Academy of Sciences–National Research Council 1956; United Nations 1958), monopolized biological effects research. Those scientists with the most specialized expertise in mutagenesis research were preoccupied with radiation. While we can acknowledge that the push to set radiation exposure limits absorbed considerable attention of those select geneticists serving on expert panels or the larger body of geneticists receiving research support from AEC contracts, it is also the case that research on chemical mutagenesis was steadily accumulating throughout this same period (Wassom 1973). In other words, the federal government's infrastructural investments in radiation genetics did not preclude progress in chemical mutagenesis. The two modes of research were in no way mutually exclusive. As I argued in the previous chapter, the local sets of resources that concentrated in radiation biology laboratories probably encouraged rather than constrained research on chemical mutagenesis.

So again, given these conditions, why did genetic toxicology arrive so late?

In posing the question in this way, I am not suggesting that genetic toxicology should have arisen earlier. The historical record is clear, however, that much of the knowledge, technology, and expertise that geneticists relied on to make their case was at least nominally available by 1955. In principle, genetic toxicology could have come about sooner. What interests me is why it didn't.

In developing an answer to that question, in this chapter I describe early efforts by geneticists to raise concerns about the public health implications of environmental mutagenesis and the resistance that thesis met from within the larger genetics community as well as from science administrators, private foundations, and industry. I also describe some of the key institutional changes during the 1960s that shifted organizational priorities and channels of influence in the federal science system that made successful mobilization around environmental mutagenesis more likely in 1970 than in 1960. Students of social movements call these shifts "political opportunities" (Kitschelt 1986; Meyer and Staggenborg 1996; Tarrow 1989) and have used the term to refer to "consistent . . . signals to social or political actors which either encourage or discourage them to use their internal resources to form social movements" (Tarrow 1996:54). As that research has shown, the fortunes of social movements can rise or fall when new governments come to power, when party alignments shift, or when relationships among political elites begin to fracture. In science, the reorganization of knowledge-production systems can similarly enhance or constrain the impacts of scientist collective action. In the case of genetic toxicology, opportunities emerging in the mid–1960s involved the expansion of existing research programs to include work on environmental mutagens, the creation of new laboratories and mechanisms to coordinate environmental mutagenesis research among those laboratories, and newfound support from political and scientific elites who lent much-needed support to the environmental mutagenesis thesis. While none of these changes caused the genetic toxicology movement or ensured its success, each shift increased the capacity for scientist collective action, either indirectly by generating broader interest in genetic toxicology or more directly by creating access to material and organizational resources not previously available to would-be scientist-activists.

Geneticists' Early Resistance to the Environmental Mutagenesis Thesis

Among those biologists involved during the 1950s in a growing national debate over the public health effects of radioactive fallout and international limits on nuclear weapons testing, there were at least a few harboring suspicions that human exposure to chemicals represented cause for additional concern. H. J. Muller and Joshua Lederberg, in particular, worried that physicists' and radiation biologists' concerns about atmospheric radiation should include the mutational effects of organic agents. These broader anxieties were premised on sound, if incomplete, mutation science. In the first of a series of letters exchanged in the spring of 1950, Lederberg remarked to Muller that "perhaps the problem is exaggerated, but I have the feeling that, in our ignorance, chemical mutagenesis poses a problem of the same magnitude as the indiscriminate use of radiations."[2] While Muller shared much of Lederberg's concern, another five years would pass before either brought those concerns to the attention of policy makers, the scientific community, or the general public. And even then, the levels of urgency they gave to the problem varied considerably. We find, for example, Muller's passing and rather vague comment, in an article on the public health hazards of radiation, that "the problem of maintaining the integrity of the genetic constitution is a much wider one than that of avoiding the irradiation of the germ cells, inasmuch as diverse other influences may play a mutagenic role equal to or greater in importance than that of radiation" (Muller 1955b:65). Far more forceful is Lederberg's (1955) pointed expression in a letter to the editors of the *Bulletin of the Atomic Scientists* that

> if we postulate survival, we cannot overlook the long-run genetic problems entirely for preoccupation with the narrower issues of public affairs. As the Bulletin shows, the attention of the informed public is rightly focused on the production of deleterious mutations by penetrating radiations, but this emphasis may have obscured the possibly wider contact of genetic hygiene with industrial civilization. . . . From this perspective, the genetic hazards of atomic energy are but one facet of a much broader and correspondingly more urgent problem of chronic toxicity and the health of the public (and its future generations). (365)

Lederberg urged that "more extensive studies are needed to establish, for example, whether the germ cells of man are physiologically insulated against such chemical insults from the environment."[3]

A similar pair of warning articles appeared in 1960, one in the British journal the *New Scientist* and the other in *Scientific American*. In this instance, however, both articles were penned by the same person—Peter Alexander, a chemist–turned–radiation biologist working on cancer therapy research at London's Chester Beatty Research Institute. In "Mutation-Producing Chemicals," Alexander (1960a) emphasized what he saw as an unjustified disparity in public and scientific concern over the dangers posed by mutagenic chemicals as compared to the widespread concern and systematic scientific and regulatory attention given to the biological effects of radiation. "During the last fifteen years," wrote Alexander, "several hundred other chemical agents have been shown to possess this property [mutagenicity], and it is certain that there are many more still to be identified." In a passage that would sound familiar to genetic toxicology scientist-activists a decade later, Alexander complained, "Nowhere has a systematic search been made, and the addition of new substances to the list is haphazard. . . . Moreover, the number of laboratories all over the world specifically engaged in the study of chemical mutagenesis is small; they are far fewer than those working on genetic effects of radiation. Almost no work has been done on the genetic effects of chemicals in mammals" (1073; see also Alexander 1960b).

These scientists' position on the topic of chemical mutagens framed a distinctly minority opinion within the postwar genetics community. Despite Muller's and Lederberg's authority as Nobel laureates and the tendency of both to use their awards as bully pulpits, early attempts to raise public and scientific awareness must have been too distant from the practical and theoretical interests of mainstream genetics to elicit much response, vocal or otherwise.[4] Viewed within the context of the radiation fallout controversy, these were modest and sporadic efforts by concerned individuals and not, as would later become the case, collective organizational responses.

Among those geneticists who considered the issue at all, most would likely have assumed that the vast majority of chemicals, synthetic or otherwise, lacked the energy potentials necessary to penetrate living cells as efficiently as radiation and therefore posed little mutagenic threat to human populations. British population geneticist J.B.S. Haldane downplayed chemical genetic hazards in his 1954 textbook, *The Biochemistry of Genetics*, by explaining that when chemical substances "are injected, or given in food, they have to pass through a number of membranes and through living substance, which destroys them to a large extent" (quoted in Goldstein 1962:167–168). This opinion was echoed by others, such as James Neel and William Schull, who wrote in their book *Human Heredity* (also published in 1954) that "because of the elaborate mechanisms which maintain the chemical constancies of the human body during life, it seems unlikely that exposure to various chemical agents exerts a very large influence on human mutation rates" (quoted in Goldstein 1962:168). Thus by 1960 there existed among geneticists both a growing interest in and enduring resistance to the thesis that chemical mutagens posed a potentially serious danger to human populations. The Second Macy Foundation Conference on Genetics, a four-day meeting held at Princeton University in the fall of 1960, provides a useful barometer of these dynamics.

The conference brought together an elite group of geneticists—Auerbach, Lederberg, Neel, and Schull among them—and was devoted mainly to group discussions centering around three presented papers on mutations and mutagenesis.[5] The first paper, "Problems of Measurement of Mutation Rates," presented by Kim Atwood (1962), a microbiologist at the University of Illinois–Urbana, dealt with the general and long-vexing problem of measuring mutation rates. The other two papers focused explicitly on chemicals. Charlotte Auerbach (1962a) gave the second paper, titled "Mutagenesis, with Particular Reference to

Chemical Factors." Her paper painted a picture of the technical state of the art, pinpointing chemical mutagens' relations to key problems in genetics theory. Avram Goldstein (1962), a pharmacologist and one of Lederberg's colleagues from Stanford University School of Medicine, gave the final paper, titled "Mutagens Currently of Potential Significance to Man and Other Species."

Goldstein's paper began with an admission that inferences about the human health risks posed by chemical mutagens remained highly constrained by a serious lack of data. Despite this uncertainty, Goldstein argued that indirect evidence raised important questions that deserved to be addressed through directed empirical research. Toward that end, he proposed a classification scheme for prioritizing chemicals for mutagenicity testing based on estimates of human exposure and described briefly those pharmaceutical drugs and other chemical substances thought to pose the greatest potential genetic hazards.[6] The paper's core centered on a detailed discussion of the toxicological, carcinogenic, and mutagenic properties of caffeine, which Goldstein (1962: 167) considered "a contender of first importance among possibly significant chemical mutagens in man."

At least some of those attending Goldstein's presentation remained unimpressed. The final exchange that day, between Sol Goodgal, a University of Pennsylvania microbiologist, and Lederberg captures the resistance demonstrated by many geneticists to the argument that the mutagenicity of exogenous chemicals posed due cause for alarm and that, by implication, geneticists bore a moral and social responsibility to pursue that line of research and to promote their findings publicly (Schull 1962:237–238):

GOODGAL: Scientists have the responsibility not to raise issues that are not based on fact. If it is demonstrated that caffeine is a mutagenic agent in a large variety of organisms, and under a variety of conditions, or if one can at least define them, then I think there is a much more solid basis for raising this issue. My own feeling is that there has been too much said today on too few facts.

LEDERBERG: The fact remains that, without generating a minimum of public excitement or at least excitement among the scientific community, nothing is going to be done about it. It is too easy to think that the results we get in the laboratory concerning mutagenesis in bacteria are merely scientific curiosities and that they can be excluded from the realm of human affairs. This is too convenient.

With hindsight, Lederberg (1997:4) has suggested that "this meeting surely helped to crystallize a scientific consensus and the beginnings of national action." Perhaps. At the very least, the dialogue initiated at Princeton and disseminated in published proceedings two years later would have given even more scientists active in genetics research increased access to the evidence and the arguments surrounding the environmental mutagenesis thesis.

Institutional Constraints to Scientist Activism

Publishing the occasional mildly polemic essay on the global and evolutionary significance of environmental chemical mutagens is one thing. Launching a concerted campaign to raise awareness about these issues that is directed at professional societies, federal agencies, international organizations, and drug and chemical companies is quite another. The stakes and the costs are considerably higher. Patronage is fundamental to discipline building, and enrolling bureaucratic organizations to give financial, material, and ideological support to emerging sciences can often mean the difference between institutional success and failure (Kohler 1991). As the 1960s wore on, the historical record provides increasing evidence that a handful of scientists were gearing up their mobilization efforts by soliciting institutional support. At the same time, it is also clear that their intensified efforts most often were met with indifference, if not active resistance, at the organizational level.

In 1962 an elderly Muller delivered a seminar to Food and Drug Administration (FDA) scientists and administrators that spelled out the possible genetic hazards of chemically induced mutations. Muller's talk seems not to have elicited much in the way of concrete responses among FDA officials who attended the seminar, his "clarion call . . . generally falling on deaf ears" (Epstein 1974:219).[7] The following year, Frits Sobels, the soon-to-be editor of *Mutation Research*, delivered an invited position paper on chemical mutagens and human safety to the World Health Organization in Geneva, where it, too, reportedly "aroused little immediate interest" (Wassom 1989:3). In the United States, Alexander Hollaender's efforts circa 1964 to promote environmental mutagenesis met early resistance from the AEC, whose research mission to study radiation effects precluded the organization of an in-house program to study environmental mutagens. As he later reminisced, "I couldn't develop [chemical mutagenesis at Oak Ridge] because we got paid for radiation work, and they didn't like it too much if we got into other areas."[8]

Thwarted at home, Hollaender took advantage of a small travel grant from the Ford Foundation to visit several research institutions, including the University of Colorado, MIT, and Children's Hospital in Boston, to no avail. "Nobody wanted to get into it. They couldn't see anything in it."[9] His efforts to convince chemical companies of the importance of mutagenicity testing through the Chemical Manufacturers' Association and the Union Carbide Corporation met with similar results, as did his efforts to convince the Genetics Society of America and the Radiation Research Society to expand their scope of professional interest into the arena of chemical mutagens and public health.[10] Coinciding with the zenith of his administrative career, Hollaender's failure to capture the attention of agency directors, corporate boards, scientific society officers, and university administrators cannot be attributed to a lack of professional status, social connections, or social influence. (As we will see in Chapter 6, the respect that Hollaender commanded among colleagues approaches legend.) Not surprisingly, others with considerably less authority than Muller, Sobels, and Hollaender found their efforts hindered from above as well. At the FDA, Marvin Legator's ultimately successful effort to establish a genetic toxicology laboratory (in 1967) was similarly conducted in the context of a "notable absence of any particular interest either inside or outside of his agency" (Epstein 1974:219). The institutional opportunities for reorganizing research on chemical mutagens remained few and far between.

One of those institutional openings, which does well to illustrate the fragile credibility on which the environmental mutagenesis thesis rested even as late as 1966, involved a one-day symposium convened at the Jackson Laboratory in Bar Harbor, Maine, on September 14. The symposium gathered fifteen members of the National Institutes of Health (NIH) Genetics Study Section (GSS) to discuss the problem of environmental mutagens. NIH sponsorship of this meeting reportedly was triggered by a letter from Harvard biochemist Matthew Meselson to President Lyndon Johnson's science adviser, Donald F. Hornig, in December 1964. The letter warned that "the prevalence of non-toxic mutagens could go unnoticed until serious damage had already been done" (quoted in Sanders 1969a:52). Discussion at the conference focused on "the general question of mutation and chemical mutagens."[11] At the meeting's conclusion, GSS chair and University of Wisconsin genetics professor James F. Crow drafted a report, with other GSS members contributing to the report's final form and content. Much of that content, according to Crow, was based on arguments and assertions delivered by Meselson during the course of the discussion.[12]

For reasons that remain unclear to Crow, NIH director James Shannon twice declined Crow's request to publish the report as an NIH document but told Crow that he should feel free to publish the report elsewhere. When asked what he thought the reasons were for Shannon's reluctance, Crow told me, "I don't have any idea. I'll make a guess—that he might be afraid that it would have committed him to some kinds of research that he didn't want to do or that he didn't think was basic or wasn't clinically related. I suspect that he felt that if this were published by NIH that it might involve some sort of commitment on his part." Two years later the article appeared in *Scientist and Citizen* as "Chemical Risk to Future Generations" (Crow 1968)—an article that I argue in a later chapter had profound implications for genetic toxicology a few years later.[13] Unlike the Macy Foundation Genetics Conference in 1960, the GSS meeting had a direct mobilizing impact. Despite institutional resistance from NIH administration—or perhaps because of it—seven of the fifteen participants became charter members of the EMS when it was established in 1969.[14]

The Changing Contexts of Environmental Health Research

In 1968 the central facts about chemical mutagens and mutagenesis remained essentially what they had been in 1960. The fundamental breakthroughs—the discovery of mustard gas mutagenesis, the description of the molecular structure, the identification of DNA as the factor of inheritance, the discovery of DNA repair processes, the development of bacterial and mammalian mutagenicity bioassays—all predated the rise of the genetic toxicology movement by at least a decade. So did the problem of extrapolation that so concerned Sol Goodgal and others at the Macy Genetics Conference in 1960. In 1968 scientists still could not say conclusively on the basis of direct evidence that chemicals found to be mutagenic in laboratory organisms posed similar risks to humans. Indeed, increased knowledge about chemical mutagen specificity and a better understanding of the metabolic differences among species (particularly among mammals) may have made the link between laboratory experiments and human genetic risk analysis more tenuous, not less. Even for those geneticists convinced that there was a problem, changes in the level of mobilizing activity between 1962 and 1968 were more of degree than of kind. If the promotional efforts of Hollaender, Sobels, and their colleagues in the United States and Europe to drum up support for research on environmental mutagenesis became more frequent during this period, they remained basically unorganized. Muller's lecture at the FDA, Meselson's letter to President Johnson's science adviser, Crow's efforts to publish the GSS report, and Hollaender's solicitations for foundation support were the individual efforts of concerned scientists, not yet the organized collective action of a scientists' social movement. The changing conditions of institutional opportunities outside the field of mutation research provide better evidence for understanding why the genetic toxicology movement emerged when it did.

Environmental Mutagenesis at Oak Ridge National Laboratory
There can be little doubt that Rachel Carson's arguments in *Silent Spring* inspired many university and government scientists engaged in pesticides research to assume a defensive posture in the subsequent investigation of the many claims she brought forth (Gottlieb 1993; Hayes 1987; Palladino 1996; Proctor 1995; Worster 1994).[15] Others, however, were inspired by more immediately visible signs of ecological harm—dead birds and squirrels on university campus grounds following herbicide application to the lawns, for instance, or changes in avian behavior noted by birding-enthusiast chemists.[16] Whatever the original source of scientists' concern, it is quite clear that in the aftermath of *Silent*

Spring, scientific institutions within the federal government became increasingly open to environmentally oriented research, with existing agencies and laboratories within the AEC and the U.S. Departments of Agriculture, the Interior, and Health, Education, and Welfare all taking steps to expand the ecological, natural resource, and environmental health dimensions of their research and/or regulatory missions.[17] The situation at Oak Ridge National Laboratory is a case in point.

In the early 1960s the AEC began to "further encourage" its laboratories to diversify their research (Johnson and Schaffer 1994:121). This policy shift was fueled in part by concern with technological overdevelopment and the questionable economic wisdom of the AEC's national laboratories remaining dependent upon a few large-scale experimental reactor programs. Over the course of the decade, AEC laboratories—but particularly Oak Ridge—began gradually to broaden their research orientation to include "environmental restoration, non-nuclear energy, and social engineering" (107). In 1967 Congress upped the AEC's ante by amending the Atomic Energy Act to place restrictions on congressional appropriations to the AEC. These budget-tightening measures virtually ensured that the national laboratories would experience budgetary short-falls. The impending fiscal crunch, which hit Oak Ridge in 1969 and extended through 1973, forced laboratory, division, and program directors to solicit external funding. In this context, the public's rising insecurity about chemicals in the environment, and the government's imperative to respond to those fears, seemed to provide a relatively straightforward and publicly salient way to ease fiscal constraints at the AEC.

Among the national laboratories, Oak Ridge led the charge. Essentially all research conducted at Oak Ridge in 1961 was paid for with AEC monies and involved nuclear science and engineering; by 1969, 14 percent of the laboratory's work was nonnuclear and derived from agencies and foundations outside the AEC, with most of this nonnuclear research oriented toward addressing environmental and health-related questions.[18] In comparison, nonnuclear research at Brookhaven, Argonne, and the other national laboratories in 1969 amounted to less than 1 percent (Johnson and Schaffer 1994:121).

Similar changes also reshaped Hollaender's Biology Division. The NIH was a major cosponsor of Biology Division research during the 1960s, first with programs to develop centrifuge and microscope technologies, then in 1965 with a National Cancer Institute (NCI) program to investigate the genetic and biochemical processes leading to carcinogenesis (Johnson and Schaffer 1994: 112–113). A central focus of the NCI-AEC "Co-carcinogenesis Program" was a study of tumor formation in mice exposed to cigarette smoke, sulfur dioxide, urban smog, and pesticides (114). This program drew heavily on Biology Division geneticists, including many of those involved in mutation research. Indeed, it was within the auspices of the Co-carcinogenesis Program that chemical

TABLE 3.1

Research Components of the NIH-AEC Carcinogenesis Program, 1965

Program Title	Research/Technical Staff
Biochemistry of carcinogenesis	7
Enzymology of the carcinogenic state	7
Mammalian chemical carcinogenesis	5
Chromosomal effects of chemicals and radiation	3
Carcinogen biology	6
Effects of radiation and chemicals on paramecium	4
Molecular biology of carcinogenesis	3
Chemical mutagenesis in microorganisms	6
Inhalation carcinogenesis	26

Source: Oak Ridge National Laboratory Biology Division, "Semiannual report for period ending July 31, 1965" (November 1965), 182. Oak Ridge National Laboratory Biology Division Archives.

mutagenesis gained an institutional foothold within the Biology Division's organizational structure (see Table 3.1). By decade's end, research on chemical mutagens was a component of work being done in several Biology Division research branches, with mammalian genetics and fungal genetics being areas of concentrated focus on chemical mutagenesis (Oak Ridge National Laboratory Biology Division 1969). Researchers in these laboratories were among the earliest to develop bioassay systems and experiments explicitly designed to gather data on the mutagenicity of chemical substances present in the human environment.[19]

The initiation in 1969 of a "[c]ooperative program on chemical mutagenesis," cosponsored by the AEC and the National Institute of General Medical Sciences (NIGMS) (Oak Ridge National Laboratory Biology Division 1970:22), and the establishment of EMIC the same year cemented the institutionalization of environmental mutagenesis at Oak Ridge. But just as the NIGMS project can be seen as a continuation of a trend throughout the 1960s of the increasing involvement of NIH monies in the research conducted in the Biology Division, the creation of EMIC needs also to be understood in a similar context.

At Oak Ridge, EMIC was "the only facility whose sole function [was] the collection, storage, and dissemination of chemical mutagenesis information," but it was not unique in its general mission, scope, orientation, or organization.[20] Oak Ridge director Alvin Weinberg began in the early 1960s to establish a net-

work of data collection centers in order to cope with the "information revolution" that, in his words, threatened to bury scientists beneath "a mound of
undigested reports, papers, meetings, and books" (quoted in Johnson and
Schaffer 1994:107). EMIC represented the seventeenth node in this larger and
already-developing network of information centers that formed the Oak Ridge
National Laboratory Information Center Complex.[21]

The point to emphasize is this: at Oak Ridge, the organization of research
on chemical mutagens, the development of mutagenicity tests, and the construction of the EMIC computer registry for the mutagenicity of chemicals did
not arise sui generis. Rather, the gradual inclusion of chemical mutagens into
the repertoire, routine, and organizational infrastructure of genetic research in
Hollaender's Biology Division was a component of a more general trend at Oak
Ridge to expand the laboratory's research horizons beyond a strict focus on
nuclear- and radiation-based science and engineering. The resulting reorganization of research within the Biology Division and its increasing focus on the
problems of environmental mutagenesis can only partially be attributed to the
impressive concentration of knowledge and technical skills of the geneticists
working there. Economic necessity also played a significant role in this expansion, as did the growing opinion of federal politicians and their science advisers that the causes and consequences of environmental pollution had become
issues of national scientific importance. Similar opportunities for mobilizing
collective action around environmental mutagenesis emerged with the creation
of entirely new institutions.

Environmental Mutagenesis at NIEHS

Established in 1969, the National Institute of Environmental Health Sciences
(NIEHS) played a central role in the early 1970s in providing an institutional
niche in biomedicine for environmental mutagenesis. As one of several newly
created research institutions established to propel scientific research on the biological and ecological effects of chemical agents, NIEHS was a key organizational
component of an emerging environmental state that also included the Environmental Protection Agency (EPA, est. 1969), the National Institute for Occupational
Safety and Health (NIOSH, est. 1970), and the National Center for Toxicological
Research (NCTR, est. 1971). In its discursive struggle for control of environmental health knowledge and policy relative to these other institutions, NIEHS
claimed a unique position.

Whereas NIOSH was concerned with "one subset of environmental health—
occupational health," NIEHS took in the "total interaction between man and
potentially toxic factors in the environment." Whereas the policy-oriented EPA
focused on the specific media in which environmental pollutants are found,
NIEHS claimed to make no such distinctions because "to understand the nature
of the compound and, subsequently, its toxicity, we study the compound both

by itself and in relation to other compounds with which it might come into contact, whether in water, food, or air." Whereas the NCTR was involved mainly in
toxicity testing and in developing standardized protocols for those tests, NIEHS
also conducted research on the underlying mechanisms of toxicity in order that
this knowledge may "eliminate the need for more and more routine testing."
And whereas other institutes within the NIH system were organized around
either specific diseases or the organs affected by disease, NIEHS is "concerned
with toxic agents regardless of the diseases they produce or the organs they
attack" (National Institute of Environmental Health Sciences 1975b:14–15). In
the institutional ecology of the environmental state, NIEHS portrayed itself as a
research center for high-quality environment-oriented health science unfet-
tered by the boundaries that the research and regulatory missions of these
other institutions reified—boundaries that distinguished modes of research
(basic, testing), categories of health (occupational, environmental, consumer),
environmental media (air, soil, water), or biological end point (toxicity, carcinogenesis, mutagenesis).

From its inception, the NIEHS was envisioned as a research complex whose
organizational structure was "open" to engagement with environmental mutagenesis.[22] The institute's mission, as stated in a 1965 report, involved nothing
less than mounting "a comprehensive attack on the environmental health
problems of the nation" (Research Triangle Institute 1965:xiii). Research undertaken at NIEHS was to "provide for the determination, study, and evaluation of
. . . the complex, inter-related phenomena underlying the human body's reaction to the increasingly wide range of chemical, physical, biological and social
environmental influences imposed by modern living" (5). As such, NIEHS
offered an institutional setting that was both ideologically and organizationally
consistent with the interests and demands of the genetic toxicology movement.

Interestingly, given the institute's charge, discussion in the 1965 report of
the technical/research components—which would include research branches
focused broadly on toxicology, physiology, pathology and cytology, and epidemiology—contained no explicit mention of a research role within any of these
organizational units for mutation research (and would not until 1973). Indeed,
only minor mention was made in the report of the potential contributions of
genetics more generally in measuring "the impact of the whole environment on
man" (Research Triangle Institute 1965:6). The report did, however, place a pre-
mium on organizational flexibility such that the structure of work "continually
encourages a multidisciplinary focus on environmental health problems." This
would be best achieved, the report continued, through a decision-making
structure that placed considerable authority among its branch chiefs for organizing the research conducted within each research branch, this to ensure that
"the details of organization [would] be governed by requirements of the
research program" (35). Accordingly, branch chiefs were empowered to allocate

resources, coordinate research, and develop mechanisms of communication and information dissemination—the very boundary-crossing activities necessary to fuel a growing interdisciplinary field.

Mutation research arrived at NIEHS late in 1972 with the formation of the Mutagenesis Branch and the appointment of Oak Ridge Biology Division geneticist Frederick de Serres to direct research toward the identification and assessment of risks posed by environmental agents to "human germinal and somatic tissue" (National Institute of Environmental Health Sciences 1972:236).[23] In 1974 de Serres's unit was renamed the Environmental Mutagenesis Branch (EMB) and under his leadership evolved rapidly into an institutional mechanism for the promotion, elaboration, and organization of environmental mutagenesis not only within NIEHS itself but also among federal agencies nationally and among environmental research and protection agencies internationally.

In addition to in-house research conducted by staff scientists, the EMB organized and supported an extensive intramural program that emphasized collaborative and contract research between the EMB and scientists at universities, private nonprofit laboratories, and other governmental agencies (National Institute of Environmental Health Sciences 1975a).[24] The EMB also pursued the vigorous development of a national-level program "to develop better coordination in environmental mutagenesis and to provide perspective for the intramural scientific staff in problem definition and resolution in the rapidly developing field." The key mechanisms for this were the monthly meetings of an "Interagency Panel on Environmental Mutagenesis," which de Serres helped to organize and chair. As a function of these meetings, EMB staff developed specific collaborative relationships with scientists from the NCI (for research on the relationship between mutagenesis and carcinogenesis of known carcinogens and noncarcinogens), the FDA (for studies of mutagenicity in food additives), and the EPA Office of Toxic Substances (for research on the mutagenicity of pesticides) and reviewed environmental mutagenesis research programs developing at other institutions, including those at Oak Ridge, Brookhaven, and Lawrence Livermore National Laboratories (152). At the international level, EMB staff members organized workshops, conferences, and collaborative research projects with Soviet, Japanese, French, and German scientists and organized training workshops in developing nations to "acquaint participants with the principal methods of mutagenicity testing" (154).[25]

Throughout the 1970s, environmental mutagenesis research and the field of genetic toxicology more generally thrived within the EMB. It was an institutional niche ideologically and organizationally consonant with the genetic toxicology movement's demands for the reorganization of genetic knowledge toward an emphasis on environmental health. In securing genetic toxicology firmly to the research infrastructure of government science, the EMB functioned as an important interdiscipline-building institution. It shifted some of

the disciplinary power away from Oak Ridge and further stabilized and legiti-
mated the emerging interdiscipline. Within the broader constellation of organizational actors busy building a newly invigorated environmental state, the
EMB also bolstered the institutional credibility of the infant NIEHS, which used
newly developed mutagenicity tests as the centerpiece of its appeal for
increased congressional funding (National Institute of Environmental Health
Sciences 1975b:2).

Elite Support

Opportunities for mobilizing collective action and for securing movement goals
often depend on direct or indirect support of political elites. It is during peri-
ods of institutional crisis or transition that elite solidarity is most likely to fracture, leading some elites to support oppositional political movements (Tarrow
1989). That support can take many forms, but in general, opportunities are created when elites are indifferent (choosing not to repress a social movement),
when elites actively repress countermovement opposition, or when elites provide direct financial, legislative, or organizational support to a social movement
(Meyer and Staggenborg 1996). Analogous to the latter scenario, elite support
provided by federal-level government officials and their science advisers in the
National Academy of Sciences (NAS) significantly influenced the reception of
the genetic toxicology movement's claims and consequent demands in toxicology and in the regulatory policy arena.

With the introduction of the Toxic Substances Control Act (TSCA) bill into
Congress in 1971, the genetic toxicology movement secured the support of three
key political figures: Senator Edmund S. Muskie (D., Maine), who chaired the
Subcommittee on Air and Water Pollution; Senator Abraham Ribicoff (D.,
Conn.), who chaired the Subcommittee on Executive Reorganization and Government Research; and William D. Ruckelshaus, the newly appointed administrator of the EPA. Even though TSCA would not be passed into law until October
11, 1976, data on chemical mutagenicity and genetic toxicology became key sell-
ing points in generating political and scientific support for the bill.

In a guest editorial in the *Forum for the Advancement of Toxicology*, Senator Muskie accentuated the need for new federal policy on the "non-obvious"
problem of chemical mutagens in the environment, noting shifting congressional interest in the long-term effects of toxic substances to which populations
are chronically exposed at low or trace concentrations. He drew on founding
EMS member Samuel Epstein's (1968) testimony before his subcommittee, and
he held up the just-created EMIC chemical mutagenicity registry as an example
of the database requirements for formulating rational policy on the use and distribution of genetically hazardous substances. Citing EMIC, Muskie (1969:1)
urged toxicologists and pharmacologists to contribute their much-needed expert-

ise in developing new techniques for studying and preventing "the mutagenic effects manifested in future generations."

Ruckelshaus was somewhat more instrumental on the occasion of a keynote address before members of the Society of Toxicology. Genetic toxicology and mutagenicity testing provided a wedge issue that Ruckelshaus could use to push the TSCA and to advocate for toxicologists' support for the EPA's regulatory authority. He told his skeptical audience, "We are on the threshold, it seems to me of a new era in experimental toxicology—one which has tremendous potential for human progress, safety, the quality of human life." His speech highlighted the "gross inadequacies" of data on the mutagenicity of trace-level environmental chemicals as both a responsibility and an opportunity. "[A] genuine effort in the study of mutagents [sic]," Ruckelshaus suggested, "could open up this whole field and produce a vast quantity of essential information in a relatively few years. You who are already doing so much valuable work in this vital field of toxicology are called upon to make your work, in every way you can, more relevant, more complete, more useful to human society in a changed and changing world."[26]

Political elites did more than simply provide rhetorical support for genetic toxicology. In some cases, their actions and influence directly generated opportunities for institutionalizing change. In 1971 Senator Ribicoff chaired hearings on "Chemicals and the Future of Man" (U.S. Senate 1971). Three EMS members— Harvard Medical School microbiologist Samuel Epstein and two NIEHS scientists, W. Gary Flamm (a microbial geneticist) and Lawrence Fishbein (an analytical chemist)—provided expert testimony at those hearings. Their remarks before the Senate subcommittee all nourished the common theme that a disturbing gap existed between current scientific knowledge about the potential genetic risks of chemical mutagens and chemical testing and monitoring requirements mandated by federal government agencies charged with protecting the public health. The data on chemical mutagenicity generated from laboratory experiments, they argued, were inconsistent with the total absence of mutagenicity testing requirements in federal policy.

Following these hearings, Ribicoff solicited from officials of the FDA, EPA, and USDA their views of the adequacy of current food safety standards. He then enlisted the EMS in evaluating those agencies' responses. "Since the EMS has given responsible and thoughtful comment on so many of these issues in the past," Ribicoff wrote in a letter to EMS president Alexander Hollaender, "I particularly look forward to your association's analysis of these three agencies' replies." At the next EMS Council meeting, Hollaender appointed a six-person committee to study the documents forwarded by Senator Ribicoff and draft a response. Perhaps not coincidentally, Hollaender also received at virtually the same time a request from the FDA-administered NCTR to nominate members to

NCTR's Science Advisory Board. As a result, four EMS members served on a committee that produced the first NCTR task force report on mutagenesis protocols.[27]

Probably more important than the efforts of any single politician in creating opportunities and providing credibility to the nascent genetic toxicology movement was support of the scientific elite embodied in the National Research Council (NRC) of the NAS. While it would be difficult to measure, there is little doubt that EMS involvement with NRC members and committees went some way toward cementing the genetic toxicology movement's credibility with governmental agencies, industry, and university biologists. Two NRC committees—the Biology and Agriculture Board and the Drug Research Board—threw their support behind the genetic toxicology movement in 1970.

A series of meetings between EMS councilors and an NRC Subcommittee on Problems of Mutagenicity were held in the spring and fall of 1970. Out of these meetings came a decision that the Drug Research Board would act as a promotional intermediary that would work to secure funding from other agencies and institutes for the EMIC mutagenicity registry by providing concerned agencies with recommendations to the EMIC funding proposals. The main thrust of these meetings, however, focused on organizing training programs in the principles and methodologies of mutagenicity testing. The purpose of the training workshops would be "to acquaint the Pharmaceutical Industry with the problem of chemical mutagenesis." The NRC committee seems to have taken an active and enthusiastic role in discussing and planning these workshops, the first of which was a symposium, "Fundamentals of Mutagenicity Testing," for about twenty senior industry and agency personnel held at the Marine Biology Laboratory in Woods Hole, Massachusetts, in the summer of 1970. This was followed in the fall by a three-day Conference on Evaluating Mutagenicity of Drugs and Other Chemical Agents for government and industry representatives held at the NAS in Washington, D.C. More than 600 were expected to attend, and the symposium received considerable attention from the press and industry trade journals, launching the genetic toxicology movement into the sphere of public debate (Schmeck 1970a,b). A more extensive and intensive "Workshop on Mutagenicity" for specialists and senior technicians was convened at Brown University in July 1971.[28]

Informal discussions between EMS councilors and members of the Biology and Agriculture Board were occurring at roughly the same time. In the spring of 1970, according to EMS minutes, this board considered "co-sponsoring a general symposium on the problems of mutagenesis and also on workshops on specific subjects of current interest and importance"; in particular, the NRC was interested "in approaching the nitrosamine problem from a broad interdisciplinary basis." The liaison in these discussions and the formal meetings that followed was Arnold Sparrow, a plant geneticist and former Oak Ridger and then-director of biology at Brookhaven National Laboratory. Sparrow served on the EMS Council from 1972 to 1975. During that time, he worked to organize an NRC workshop on environmental mutagenesis and was appointed head of an EMS subcommittee to pursue those plans.[29]

Although NRC support was not automatic, both the Drug Research Board and the Biology and Agriculture Board provided much-needed organizational and financial support during the genetic toxicology movement's infancy.[30] In its broader gate-keeping function, the NAS furnished a mantle of authority and credibility to the claims of EMS scientists and, through these NRC committees, provided institutional linkages between the movement's demands and the policy concerns of the federal government.

The support from political and scientific elites in this case is no mystery, much of it being generated through personal connections that fed mutual interests. Hollaender's membership in the Washington-based Cosmos Club (noted in Chapter 1), Meselson's work on chemical and biological weapons policy, Epstein's consulting role in the Senate Subcommittee on Public Works (1970–1974), his congressional testimony on the health effects of several environmental chemicals, and his participation in drafting the original version of the TSCA in 1970 gave these men ready access to high-ranking officials in Congress and the executive branch.[31]

The social connections between the EMS and the NAS are even more in evidence. These connections depended far less on direct advocacy (Epstein, Meselson) or membership in a cultural elite (Hollaender) than on overlapping memberships: several EMS members belonged to the NAS. Arnold Sparrow, who advocated for the EMS from within the Biology and Agriculture Board, was one. Also included in this group were George Cosmides and Paul Calabrisi, two pharmacologists who both served simultaneously as EMS councilors and academy members serving on the NRC Subcommittee on Problems of Mutagenicity. Alexander Hollaender was another, as were a number of other geneticists— among them, Joshua Lederberg, James V. Neel, and Oak Ridge mouse geneticist William Russell. The overlapping memberships of the EMS and the NAS suggests that, at its core, the genetic toxicology movement was firmly connected to some of the more elite echelons of science and government.

Conclusion

While most of the theoretical and technical machinery to bring about genetic toxicology was in place well before 1969, the organizational machinery was not. Throughout the 1950s and early 1960s, the environmental mutagenesis thesis consistently met indifference and sometimes active resistance from colleagues, science administrators, foundations, and industry. Scientist advocacy was grounded in the efforts of concerned individuals, not in organized collective action. The opportunity structures embedding environmental mutagenesis research remained highly constrained.

Beginning in the mid–1960s, institutional changes in the organization and administration of federal science and congressional interest in environmental pollution created new opportunities for coordinated and focused research on the genetic effects of environmental chemicals that in turn encouraged scientist collective action. The expansion of existing research programs to include environmental health, as occurred in the Biology Division at Oak Ridge National Laboratory, and the creation of new research centers, such as NIEHS, quite literally made room for studies of "environmental mutagenesis." Support for genetic toxicology research from national political elites like Senators Muskie and Ribicoff and from elite organizations such as the NAS connected the problem of environmental mutagens to power structures in government and science policy. These changes in the organization of research and in the channels of political influence helped to generate broader interest in genetic toxicology and gave would-be scientist-activists access to material and organizational resources not previously available. The absence of these conditions constrained scientist activism prior to 1969, just as their presence facilitated the mobilization of scientist activism thereafter. Another way of putting the argument is that while some resistance to genetic toxicology was always present, after 1969 the institutional resources to overcome that resistance help explain why genetic toxicology arose when it did.

But if movements succeed most when they take advantage of emerging opportunities, students of social movements also recognize that movements "create opportunities for themselves and others . . . by diffusing collective action through social networks and by forming coalitions of social actors; by creating political space for kindred movements and countermovements; and by creating incentives for elites to respond" (Tarrow 1994:82). As we'll see in later chapters, the scientists' movement that built genetic toxicology not only seized opportunities created by transformations in the organizational structure of environmental science but also created opportunities of its own. Scientists collectively generated opportunities by strategically framing the health risks associated with environmental mutagens, creating new organizations, and developing public outreach and education programs. They also consciously included industry scientists in the movement, thereby altering traditional divisions and alliances among scientists working in different research sectors and guided by different professional interests. We'll gain a better appreciation for the importance of these opportunity-creating efforts if we have a clearer picture of the character and scope of scientist collective action that burst forth in 1969.

4

A Wave of Scientist Collective Action

> The barricade construction moving beyond neighborhoods in the French
> revolution of 1848, the factory councils in the Russian revolution of 1905,
> the sitdown strikes of the French Popular Front and the American New
> Deal, the "direct actions" of the 1968–1972 period: in the uncertainty and
> exuberance of the early period of a cycle of mobilization, innovation
> accelerates and new forms of contention are developed and diffused.
>
> –Sidney Tarrow, *Power in Movement*

Beginnings of social movements are notoriously difficult to pin down, the origins of the genetic toxicology movement no less than others. Institutionalizing efforts date at least to Alexander Hollaender's extended tour to biology and genetics laboratories in Europe during the summer of 1967. Having just stepped down from his post as director of the Oak Ridge National Laboratory Biology Division, Hollaender put "[c]onsiderable effort . . . into trying to convince pharmacologists, as well as other people, of the importance of chemical mutagenesis." His travel report noted that chemical mutagenesis "could very well become a kind of focal point for further development of cooperation with different government agencies in building up new approaches to basic biology using the same type of approach which has been used in the radiation field. These projects are still under discussion" (Oak Ridge National Laboratory Biology Division June 1968:78–79). Back home, Hollaender and a few other Oak Ridge biologists began to lay the institutional groundwork for genetic toxicology. In April 1968 they sponsored an "informal discussion" on mutagenesis at the Oak Ridge Biology Division among interested division members and three "outside investigators." The next September, a "Roundtable on Mutagenesis" in Gaithersburg, Maryland, attracted forty biologists from around the country for informal discussion of practical tests for mutagenicity, monitoring human populations for increases in the load of mutations, and "necessary emphases for future research," among other related topics (Oak Ridge National Laboratory Biology Division 1969:50). At this meeting, the decision was made to form a professional society to address these concerns.[1] Eight months later, articles of incorporation were signed that established the EMS as a tax-exempt scientific organization. As expressed in an announcement published in *Science*, *Nature*, *Genetics*, and several other major science journals, scientists established the EMS "to encourage interest in and study of mutagens in the human environment, particularly as these may be of concern to public health."[2]

With the establishment of a formal organization, the project to establish genetic toxicology began in earnest. During the next five to seven years, a sustained flurry of interdiscipline-building activity included the establishment of laboratories, journals, annual meetings, and funding mechanisms, as well as large-scale collaborative interlaboratory and interagency research programs. My main focus in this and the following two chapters is on genetic toxicology's initial rapid development, from 1968 to roughly 1976. This was a period of intensive institutionalization, when much of genetic toxicology's social structure was established and when community identity among geneticists, biochemists, and toxicologists first coalesced into something new. It was also the period in which the advocacy work of individuals began to take new forms as purposeful collective action; when a loose set of ideas became formalized into strategy and tactics; and when a small group of concerned colleagues became activists at the center of a growing scientists' movement.

I mark this transformation analytically, empirically, and rhetorically. Earlier chapters have emphasized the social contexts of mutation research, the institutions that structured knowledge about chemical mutagenesis, and the role of chemical mutagens in the political economy of genetics practice. The rest of the book examines the scientists' social movement that created genetic toxicology. From here on, the analysis is guided more by social movement theory than by historical science studies, and my earlier concern for historical detail gives way to more general theorizing about the processes that constitute scientists' social movements. My empirical focus also shifts in degrees, if not in kind, from research to politics, from individuals to organizations, from advocacy to collective action. Until now, I have described the main actors—Muller, Auerbach, Lederberg, Crow, Meselson, and others—as advocates, promoters, entrepreneurs, or simply concerned individuals. I use a different language now. The term "scientist-activists" describes those individuals who organized and participated in the movement to create genetic toxicology. Underlying this term is my understanding that scientist activism in genetic toxicology was organized, strategic, and collective action that was guided at least as much by environmental values and by a social critique of the limitations of discipline-based science as it was by the disciplinary values that had previously influenced mutation research.

I interpret this movement as contentious politics, but it was a subtle form of contention, primarily implicit in scientist-activists' conventional institution-building activities. The movement challenged the disciplinary organization of professional science, federal environmental policy, the basic/applied dichotomy distinguishing biology and public health, and the social responsibility of geneticists and toxicologists. To make these challenges credible, scientist-activists drew on forms of collective action that were at once scientific and political, for example, by conducting experiments on genetic effects of artificial preservatives, by publishing commentary in *Science* calling for bans on mutagenic substances, or by creating professional societies that sponsored conferences but also engaged in public education and outreach. In other words, scientists used thoroughly routinized mechanisms built on research, communication, and collaboration to build an unconventional science. How this happened are topics for Chapters 5 and 6.

The present chapter presents a bird's-eye view of the "wave" of movement activity attendant to the rise of genetic toxicology. I consider three distinct types of aggregated data: the scientific literature on chemically induced mutagenesis, organizational phenomena such as conferences and technical workshops, and the social and professional characteristics of EMS members as a group. Together, these data provide provisional answers to when and where scientist activism happened, what forms that activism took, and which scientists were most centrally involved. In so doing, this chapter lays the empirical foundation for examining in detail the further questions of how scientist collective action was organized and why it took the forms it did.

A Research Production Boom

After nearly three decades of nominal increase, in 1968 published research on chemical mutagens and mutagenesis began to skyrocket. A descriptive study of this literature noted that "since 1968, published material [on chemical mutagens/mutagenesis] has grown at a 200–500 rate of increase per year" and estimated that approximately 2,500 papers were published during 1972 alone (Wassom 1973:276). Of the 6,094 citations in the EMIC database for the years 1968–1972, 81 percent (n = 4,957) contained original data published in scientific journals. The study showed that 109 of these journals published ten or more of these citations, suggesting that scientific interest in chemically induced mutagenesis was widespread. The range of disciplinary and thematic foci represented by such journals as *Food and Cosmetic Toxicology*, *American Journal of Botany*, and the *New England Journal of Medicine* illustrates that chemical mutagenesis research bore implications far beyond the theories and experimental practice in classical genetics. Moreover, the number of citations in such prestigious journals as *Proceedings of the National Academy of Science* (101), *Nature* (164), and *Science* (106) suggests that editors of these journals considered chemical mutagenesis a topic of some significance (Table 4.1).

The preeminent outlet for data derived from chemical mutagenesis

TABLE 4.1

Top Ten Journals for Chemical Mutagenesis Information
(cumulative through 1972)

Publication Source	Number of Citations
Mutation Research	410
Genetika (USSR)	251
Genetics (U.S.)	248
Nature	164
Journal of Bacteriology	136
Cancer Research	130
Molecular and General Genetics	127
Science	108
Environmental Mutagen Society Newsletter	106
Proceedings of the National Academy of Science (U.S.)	101

Source: John S. Wasson, "The literature of chemical mutagenesis." In Chemical Mutagens: Principles and Methods for Their Development, vol. 3, edited by Alexander Hollaender, 271–287. New York: Plenum Press.

experiments, however, was the journal *Mutation Research*. A closer examination of its pages provides additional detail of the transformations that characterized the production boom (Table 4.2). The journal's subject index for 1968 shows 35 listings for chemical mutagens; in 1972 there were 361. This represents a more than ten-fold rate of increase in just five years.[3] In comparison, index listings for ultraviolet, gamma, and x radiations also rose during that same period, from 45 to 84, but the increase was far less in both absolute and relative terms. The point is even more starkly made when we control for the variation in number of volumes published (two in 1968, three in 1972) and the differences are calculated as a percent of change. Index listings for radiation increased 24 percent; index listings for chemical mutagens increased 588 percent. These differences suggest quite clearly that growth in the field of mutation research in the late–1960s and early 1970s can be attributed largely to rising research interest in chemical mutagens.

Institutionalizing Events

It is common practice among students of collective behavior to measure cycles of social protest by counting disruptive events (e.g., marches, sit-ins, riots)

TABLE 4.2

Listings for Chemical and Radiation Mutagenesis in
Mutation Research Subject Index, 1968 and 1972

| | 1968 | | | 1972 | | | | |
	v. 5	*v. 6*	*Total (ave.)*	*v. 14*	*v. 15*	*v. 16*	*Total (ave.)*	*% Change**
Chemical	27	8	35 (17.5)	131	78	152	361 (120.3)	59
Radiation	20	25	45 (22.5)	24	33	27	84 (28.0)	24
No. of articles	48	51	99	55	51	54	160	

Source: Mutation Research vols. 5, 6, 14–16 (1968 and 1972).

* Percent change (x_2-x_1/x_1) was calculated using the yearly average of index listings to control for variation in the number of volumes published in 1968 and 1972.

described in newspaper reports or in police, prison, or other public records, and then aggregating these events over time to gain an understanding of where, when, and in what manner social protest occurred (Oliver and Myers 1999).[4] In similar fashion, a "picture" of the wave of scientist collective action that defined the movement to establish genetic toxicology begins to emerge by aggregating "institutionalizing events." I define these as any collective event relevant to or promoting the scientific investigation of chemically induced mutagenesis, mutagenicity test development, or genetic hazard identification. These include but are not limited to conferences, symposia, training workshops, and the formation of professional organizations. A careful examination of announcements, news briefs, and conference reports contained in journals, society newsletters, and laboratory progress reports reveals a sustained concentration of institutionalizing events after 1969 (Figure 4.1).[5] Before that time, institutionalizing events promoting genetic toxicology–relevant research were sporadic, averaging less than two per year during 1964–1968. However, in 1969 the rate of institutionalizing events jumped to twelve, beginning a sustained upward trend and averaging thirteen per year through 1976.

As a collective phenomenon, institution building in genetic toxicology was characterized by the temporal concentration of events. Where once the genetic hazards of environmental chemicals had been a topic deserving comment by an invited speaker or a single panel discussion, after about 1970 entire conferences, symposia, and training workshops devoted to the subject became commonplace. Concentrated in time, institutionalizing events were dispersed in

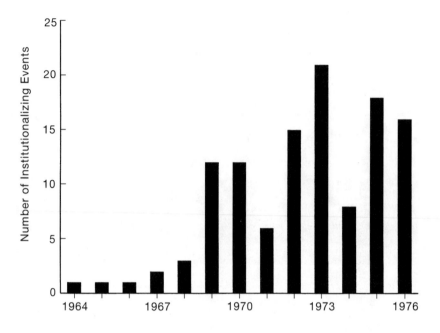

FIGURE 4.1 Institutionalizing Events in Genetic Toxicology, 1964–1976
Sources: See Appendix B.

both intellectual and geographical space. As will become clear, the movement
was both interdisciplinary and international. As just one indicator, the literature on chemical mutagenesis published during the period 1968–1972 could be
found in "approximately 700 sources" (Wassom 1973:278). Within a very short
period of time, researchers from many different countries, disciplinary backgrounds, and specialty interests began to meet to discuss the research, methodology, and policy issues surrounding genetic toxicology.

Other qualitative changes involved the research communication infrastructure that in part developed out of these face-to-face exchanges. Important
outcomes included the publication of three textbooks on mutation and mutagenesis (Auerbach 1976; Drake and Koch 1976; Drake 1970) and the completion
of five volumes of what would eventually become a ten-volume monograph
series on methodologies for detecting chemical mutagens.[6] But the point is particularly acute in the case of regularly published, peer-reviewed journals: in the
early 1960s no formal publication outlet specifically devoted to the study of
chemical environmental mutagenesis existed (Sobels 1964). Although the creation of *Mutation Research* in 1964 began to address this need, by 1976 that
journal had expanded to include three new sections, each published separately.
The first section, *Environmental Mutagenesis and Related Subjects*, began in
1973. Research published here focused mainly on test development and proto-

col standardization. Two more sections—*Reviews in Genetic Toxicology*, which published in-depth critical assessments of the mutagenic potential of various chemical agents, and *Genetic Toxicology Testing*, which provided an outlet for positive and negative mutagenicity testing results—were begun in 1975 and 1976, respectively. These new scientific journals helped put genetic toxicology on the scientific map by organizing the rational production of new knowledge on environmental chemical mutagens.

A topic previously of note mainly to geneticists, biochemists, and agricultural breeders, during the 1968–1976 period chemical mutagenesis would become reconstructed as a public health problem and the basis of an emerging environmental health science. The wave of institutionalizing events that mirrored this conceptual transformation also reflected the development of an international research-training-communication infrastructure. Before examining how these new arrangements were achieved, it is useful to consider the group characteristics of the scientist-activists involved in this interdisciplinary project.

The Scientist-Activists

Who were these scientist-activists who stepped outside their daily routines as research scientists and administrators to promote genetic toxicology and to raise public awareness of the genetic hazards of environmental chemicals? From what research communities did they come? What social characteristics, if any, did they share? We can begin to shed some light on these questions by examining the social and professional characteristics of those people who in 1969 became members of the EMS.

Professional societies in science most often arise to provide organizational support and institutional legitimacy to scientific specialties (Whitley 1974). Membership in professional societies, correspondingly, represents some core component of the general population of research scientists working in any particular specialty area. Since the EMS at the end of its first year constituted the only professional science organization specifically promoting research on the human genetic effects of environmental chemicals, membership in the EMS serves as a close proxy of those scientists working in 1969 who held an abiding research interest in that topic. The first EMS membership list available (from June 1969) contains only 87 members. Over the next six months, however, total membership in EMS nearly tripled. Another list, generated in December 1969, contained information on 276 members. It provides a clearer picture of the people initially motivated to join that society (Table 4.3).

By the end of its first year, general membership in the EMS in many respects resembled what might be considered "typical" of professional biological societies of the day in that the vast majority of EMS members were men,

TABLE 4.3

Environmental Mutagen Society Membership by Institutional Affiliation and Other Social Characteristics, December 1969

Federal research institutions	57	
Atomic Energy Commission laboratories and divisions[1]		27
National Institutes of Health		9
Food and Drug Administration		5
U.S. Department of Agriculture		3
Other U.S.[2]		3
Foreign (non-U.S.)		10
Academic research institutions[3]	119	
Biomedical sciences		51
Life sciences		66
Agricultural sciences		2
Private research institutions	53	
Nonprofit laboratories		9
Industry laboratories		44
Public interest organizations[4]	2	
Federal government[5]	4	
State and county government[6]	2	
Number of women members	24	
Members not identified by institutional affiliation or identified as nonscientists	33	
Members identified as holding an administrative role	55	
Non-U.S. members[7]	40	

Source: Total membership = 276 "Membership list," *EMS Newsletter* 2 (1969): 74–85.
Notes:

[1]Thirteen AEC-employed members represented Oak Ridge National Laboratory.
[2]Consumer Protection and Environmental Health Services (1), Environmental Control Administration (1), National Science Foundation Cell Biology Division (1).
[3]My designation of academic departments, programs, units, or laboratories as belonging to one of these three broader administrative divisions was not systematic. Some coding decisions were based on knowledge I had about certain individual's specific lines of research, but most were based on best guesses given the information at hand—essentially member's mailing addresses. On the other hand, there is considerable blurring across the broad spectrum of the biological sciences, and some measure of arbitrariness is to be expected.
[4]Scientists' Institute for Public Information (1), Environmental Information Center (1).
[5]U.S. Public Health Services (1), U.S. Office of Standards, Consumer Protection, and Environmental Health Services (1), U.S. Senate Committee on Public Works (1), Department of Health and Social Security (England) (1).
[6]California Dept. of Public Health, Dept. of Public Health (Detroit and Wayne Counties, Mich.).
[7]W. Europe (29), E. Europe (3), Australia (1), Canada (4), India (1), Israel (1), Philippines (1).

most of whom held positions in academic research institutions. Researchers from government laboratories were also heavily represented in EMS, with employees of AEC laboratories and divisions contributing a disproportionate number. In academia, the members' disciplinary affiliations cut across the life, biomedical, and agricultural sciences. The same holds for the relative proportion of members employed at private nonprofit research foundations. Interestingly, even though there were no for-profit genetic toxicology laboratories in the United States until the early 1970s and federal regulations would not require mutagenicity testing of new chemical compounds until 1976, industrial laboratories contributed 16 percent of the EMS membership in 1969. Their presence was in part a reflection of an early decision by the charter members of EMS to actively recruit representatives from industry and in part an acknowledgment by pharmaceutical or chemical laboratories that their economic and political interests would be best served by becoming involved in mutagenicity research; promoting genetic toxicology provided these firms a means of gaining ground-floor access to the regulatory decisions that were likely to flow from the FDA and EPA.[7]

Forty foreign scientists also joined the EMS during its inaugural year. Their presence suggests that scientists' concern over environmental mutagens had strong international resonance despite cross-national differences in the structure of environmental regulation and science policy and the relative strength or weakness of national environmental movements. These international networks were important in the rapid creation of several EMS societies in Europe and Asia in the early 1970s. The variety of EMS members' disciplinary and departmental affiliations, clustered in the life and biomedical sciences, is even more impressive (Table 4.4). The thirty-one different departments or disciplines identified by members' institutional addresses ranged from entomology and food science in agriculture to oncology and pediatrics in the health sciences and from botany to molecular biology to zoology in the basic life sciences. Admittedly, the nomenclature for these types of intra-institutional divisions is somewhat arbitrary and does not necessarily reflect the substantive content of the research conducted within them. Nevertheless, the apparent proliferation of organizational labels does suggest that a significant transformation was under way with respect to which scientists in 1969 were paying attention to chemical mutagenesis. If nothing else, it is clear that the research topics that would become the central problematics of genetic toxicology were beginning to attract the attention of scientists in fields of research previously little concerned with the genetic impacts of chemicals.

For a small and fledgling professional society, the EMS was long on relatively powerful members—not just those scientists who had distinguished track records in basic genetics but also those in science policy, funding, regulation, and administration with access to decision makers in universities and in fed-

TABLE 4.4

EMS Membership Self-Identified by Department and/or Discipline, December 1969

Life Sciences	Health Sciences	Agricultural Sciences
Biology (27)	Pediatrics (6)	Entomology (1)
Genetics (16)	Environmental medicine (6)	Food science (1)
Biochemistry (7)	Medical genetics (3)	
Zoology (7)	Oncology (3)	
Radiation biology (4)	Mutagenesis (3)	
Microbiology (4)	Health physics (2)	
Botany (2)	Pathology (2)	
Chemistry (2)	Toxicology (2)	
Cytogenetics (2)	Biophysics (1)	
Radiation genetics (2)	Chronic disease (1)	
Cell biology (1)	Dermatology (1)	
Molecular biology (I)	Environmental health (1)	
	Health science (I)	
	Medicine (1)	
	Nutrition (1)	
	Pharmacology (1)	
	Radiation health (I)	

Source: "Membership list," EMS Newsletter 2 (1969): 74–85.

eral government. In all, fifty-five EMS members held some sort of administrative position. These included heads or chiefs of federal agency laboratories, department chairs, office or project directors, two university deans, and the vice president of research for a pharmaceutical company. In addition, a few EMS members worked in the regulatory or health science policy arenas. Among this group, we count an executive secretary from the NIH Genetics Study Section, a member of the Cell Biology Division of the National Science Foundation, a staff member of the U.S. Senate Subcommittee on Public Works, representatives from two pre-EPA-era environmental health science bureaus, two representatives from state government agencies (California and Michigan), and the U.S. Public Health Service assistant surgeon general. With bureaucratic authority to either directly or indirectly influence the organization of research in their own laboratories, departments, or divisions and with some access to other decision makers higher up in the academic, governmental, or corporate hierarchy, these

EMS members represented a potentially influential force for change within the institutions of science.

Some other members, of course, presumably enjoyed considerably less social status and institutional authority. Among that group would likely have been many of the thirty-three people not identified by institutional affiliation or occupational title. Since it is likely that most members who did have a direct research or administrative relationship to environmental mutagenesis noted this on their membership forms, it is fairly safe to assume that those who omitted such information did not have a direct or current professional relationship to the field. Some, like Karl Sax, were retired scientists.[8] Others, like Frank Di Luzio, were entrepreneurs who had prior experience in regulatory and policy arenas.[9] Two others were representatives of public interest organizations. Since eligibility for membership in the EMS was open to "scientists and others," it is probably the case that most of those members whose mailing address lacks any institutional or occupational affiliations were not themselves practicing scientists.[10] This theory is supported by the fact that the formation of the EMS received considerable publicity not only in scientific journals but in trade journals and newspapers as well. Also, twelve of these addresses are clustered in the Washington, D.C., area, where the first EMS annual meeting was held. Another smaller cluster is located in and around Oak Ridge, Tennessee. It is not clear how long these people remained EMS members, but it seems fairly certain that there was at least some level of initial involvement from people other than active research scientists and science administrators. In its origins, the EMS reflected something more than a theory-driven intellectual movement and functioned as something more than a professional scientific society. Its membership cut across sectoral and disciplinary divisions and was nominally open to nonscientists.

It is important, however, to distinguish between an organization's general membership and those who function in leadership capacities and as spokespeople for the organization. Indeed, among this latter group may be found most of the committed scientist-activists whose efforts first gave meaning and movement to the field of genetic toxicology. In August 1970 the leadership of the EMS was drawn exclusively from the organization's charter members—those scientists who attended the January 1969 organizational meeting and formally voted the EMS into existence. To best understand the balance of power within the EMS and the genetic toxicology movement more generally, we need to look more closely at these central actors.

The Activist Core

When viewed in the aggregate, it becomes clear that the small core of scientist-activists who created and presided over the EMS in its first years as officers and councilors was considerably less heterogeneous than the EMS as a whole (Table 4.5).

TABLE 4.5

EMS Officers and Councilors, August 1970

Name	Age	Institution	Position	Discipline/s	Specialty Area/s[1]
A. Hollaender[2,3] (president)	72	ORNL, Biology Div.	Sr. Research Adviser	Biophysics	Effects of ultraviolet on bacteria; proteins & nucleic acids
M. Messelson[2,3] (vice pres.)	46	Harvard, Biology Dept.	Professor	Molecular Biology	Biochemistry and molecular biology of nucleic acids
M. S. Legator[2,3] (treasurer)	44	FDA	Chief, Cell Biology Branch	Biochemistry, Bacteriology	Mutagenicity studies; Microbial genetics
S. S. Epstein[2,3] (secretary)	43	Children's Cancer Research Found. (Boston)	Chief, Labs of Carcinogenesis & Toxicology	Pathology, Environmental Sciences	Toxicology; carcinogenesis; mutagenesis
E. Freese[2,3] (president, ex officio)	44	National Institute of Neurological Diseases and Stroke	Chief, Lab. of Molecular Biology	Biology	Molecular mechanisms of mutation
J. F. Crow[3] (vice pres. ex officio)	54	U. Wisconsin, Genetics and Medical Genetics	Professor & Chair	Genetics	*Drosophila* & population genetics
E. H. Y. Chu[2]	43	ORNL, Biology Div.	Biologist	Genetics	Mammalian cytogenetics; somatic cell genetics
G. Cosmides	44	National Institute of General Medicine	Director, Pharmacology-Toxicology Program	Pharmacology	Drug metabolism; pharmacogenetics

Name	Age	Institution	Position	Discipline/s	Specialty Area/s[1]
F. J. de Serres[3]	40	ORNL, Biology Div.	Coordinator, Environmental Mutagenesis Program	Microbial Genetics	Radiation; chemical and environmental mutagenesis; mutagenicity of carcinogens
J. W. Drake[2]	38	U. Illinois U-C, Microbiology Dept.	Professor	Genetics, Virology	Replication and genetics of bacterial viruses; molecular mechanisms of mutation
C. W. Edington	45	AEC, Div. of Biology & Medicine	Chief, Biology Branch	Genetics	Radiation biology
L. Friedman[2]	55	FDA, Bureau of Foods	Director, Div. of Toxicology	Food Toxicology, Nutrition	Biological assay; experimental carcinogenesis; evaluation of food safety
H. B. Gelboin	41	National Cancer Institute, Etiology	Chief, Chemical Branch	Biochemistry	Biochemical mechanisms of carcinogenesis
J. J. Hanlon	58	U.S. Public Health Services	Assistant Surgeon General	Public Health	Public administration in preventative medicine & public health
K. Hirschhorn[2]	44	Mt. Sinai School of Medicine	Professor	Internal Medicine	Human genetics; immunogenetics
H. Kalter[2]	46	U. Cincinnati, College of Medicine, Pediatrics Dept.	Assoc. Professor	Genetics, Teratology	Experimental mammalian teratology
C. J. Kensler[2]	55	Arthur D. Little, Inc.	Sr. Vice President in Charge, Life Sciences Div.	Pharmacology, Biochemistry	Nutrition & cancer; tissue metabolism; mode of action of carcinogenic agents; industrial toxicology
D. H. K. Lee	65	National Institute of Environmental Health Sciences	Associate Director	Physiology	Climatic physiology

continued

Name	Age	Institution	Position	Discipline/s	Specialty Area/s[1]
H. J. Malling[3]	39	ORNL, Biology Div.	Research Staff	Genetics, Microbiology	Mutation induction; enzyme production by microorganisms; induction of cancer
J. V. Neel	65	U. Michigan, Human Genetics Dept.	Professor and Chair	Population Genetics	Genetics of humans
L. E. Orgel	43	Salk Institute of Biological Studies	Senior Fellow	Chemistry	Transitional-metal & prebiotic chemistry
W. L. Russel	60	ORNL, Biology Div.	Principal Geneticist	Genetics	Genetic effects of radiation; genetics of the house mouse
J. Shubert		U. Pittsburgh, Graduate School of Public Health			
P. Shubik	49	U. Nebraska College of Medicine, Epply Institute	Director and Professor	Pathology, Oncology	Chemical carcinogenesis; environmental and industrial cancer; toxicology

Source: American Men and Women of Science Vols. 1–6 (Physical and Biological Sciences), 12th edition. New York: R. R. Bowker (1971).

Notes:

[1]In cases where individuals listed several research interests, I chose those two or three that seemed most closely related to environmental mutagenesis.

[2]Attended first EMS meeting in Washington, D.C., February 8, 1969.

[3]Ad Hoc Committee of the EMS, New York City, January 8, 1969. James Crow and Bruce Ames invited but absent.

The first thing to note is that the age structure of this group in August 1970 was strikingly concentrated, with 70 percent between the ages of forty and fifty-five. Presumably, most had reached a stage of their careers where they enjoyed a measure of reputational authority and job security. All those who worked in academia had tenure, and all but one was full professor; two were department chairs. All but two of those employed in government or private laboratories occupied administrative as well as research positions, for example, as laboratory director, branch chief, or division head.

The institutional locations and disciplinary and research interests of these core scientist-activists also cluster differently than among the general membership. Most EMS members identified by research institution were affiliated with universities, medical or public health schools, or hospitals (about 53 percent), with affiliations in government and the private sector falling behind at roughly 24 percent and 23 percent, respectively. The institutional affiliations of the officers and councilors of the EMS are distributed differently, with thirteen of the twenty-four representing government institutions (54 percent); eight representing universities, medical schools, and training hospitals (33 percent); and only three representing the private sector (13 percent). Importantly, chemical or pharmaceutical producers were not represented at all in the EMS leadership. Two members represented nonprofit research foundations, and one represented a for-profit business consulting firm. Regarding disciplines and research specialties, the core was clustered mainly in genetics and in those other basic biological sciences dealing with subcellular processes and, for the most part, submammalian systems. The biomedical sciences are underrepresented relative to the general body of the EMS. Consisting mostly of midcareer geneticists with accomplished research records in university or government laboratories, along with a few late-career science administrators, as a group the officers and councilors of the EMS were firmly established in the institutional mainstream of American biological science. All had accomplished research careers, and their reputations and jobs were secure.

Some General Inferences

This descriptive analysis provides a baseline understanding of the genetic toxicologists' movement. It begins to answer questions about when and where the movement occurred, what kinds of collective action the movement involved, and who were the movement's core activists. The most intense wave of scientist collective action began around 1968, consonant with a proliferation of scientific literature on chemical mutagenesis. The formation of the EMS in 1969 marks the beginning of a parallel wave of institutionalizing events that signal both quantitative and qualitative changes in interdiscipline-building activity in genetic toxicology.

Although this brief analysis of the social and professional characteristics of EMS leaders and its general membership says relatively little about how they organized and successfully managed a movement to establish genetic toxicology as a public health science, we do know that the EMS represented a group of people that was heterogeneous along a number of dimensions. Most members were career scientists, but a significant minority were not professional researchers. Ten percent did not reside in the United States. Most scientists worked in academic settings and represented a mix of agricultural, biomedical, and basic life sciences. Government laboratories were also highly represented.

The EMS leadership was significantly less diverse: most officers and council members worked in government institutions; nearly all had formal training in genetics; and, with the exception of a few older science administrators, nearly all were midcareer scientists who enjoyed relatively high status among their peers and secure employment. The leadership, in short, represented an elite core of mainstream life scientists. As a social movement organization, the EMS benefited from the heterogeneity of the general membership and from the homogeneity of its officer/councilor core. As we will see, both served as important resources for the movement.

This analysis also suggests something about what the genetic toxicology movement was not. We learn, for example, that the rise of genetic toxicology was not primarily brought about by the professional maturation of a new generation of scientists. In general, the most committed scientist-activists were not newly minted Ph.D.s fresh from graduate school, politicized by campus life, and eager to foment change in their chosen profession. Unlike so many of the radical professional movements of urban planners, engineers, health workers, or public interest lawyers that emerged in the 1960s and 1970s, the genetic toxicology movement was not, primarily, a student-based or young professionals' movement (Hoffman 1989; Layton 1971; Pandora 1997).

Collective action research suggests that people choose to engage in social protest when the risks are relatively low and when they perceive that there is some chance of being successful (Tarrow 1994). Status and power tend to come at a later career stage for most scientists than for other professionals and nonprofessionals; scientists also tend to begin their professional careers later in life, reaching professional maturity in their fifties. As one's position and status become secured as a result of tenure or establishing a solid record of research, activism becomes both relatively less risky and more likely to result in success. Thus it should not surprise that most of the key organizers of the genetic toxicologists' movement were midcareer professionals.

We also can infer that the key social networks connecting scientists active in the genetic toxicology movement were disciplinary and institutional, not pedagogical. Most of the core activists in EMS were formally trained geneticists, and most were connected to government research institutions either as employees, as consultants, or as members of expert review committees. Thus this rise of genetic toxicology should not mainly be attributed to the growth and dispersion of intellectual family trees; this is not, in other words, a genealogical story. Of course, intergenerational networks connecting mentors to their students and intellectual "grandchildren" did play some role in the development of genetic toxicology, as such relations do in all scientific fields (Kohler 1994; Mullins 1976). But, as we will see in later chapters, these were not the primary networks that initially tied the core scientist-activists together in 1969.

5

Framing Scientist Activism

> Of all the environmental pollutants to which we are exposed, probably
> the most dangerous from both the immediate and especially the long-
> term point of view are those which could cause changes of the genetic
> make-up–mutations in both germinal and somatic cells.
>
> —Marvin Legator, untitled manuscript

In addition to the quantitative increase in research productivity, the institutionalization of genetic toxicology also involved a qualitative transformation in
scientists' perceptions of what mutagenic chemicals were. Scientific discourse
surrounding chemical mutagenesis itself began to mutate around 1968. In June
a revised version of the NIH GSS report on chemical mutagens described in
Chapter 3 appeared in the public interest journal *Scientist and Citizen* as an
article titled "Chemical Risk to Future Generations" (henceforward "Chemical
Risk"). Written by GSS chair James F. Crow (1968), the article explained in lay
terms geneticists' major concerns regarding the genetic risk of human exposure
to chemical mutagens and outlined a program of action to minimize the potential danger.

In the years immediately following the article's publication, Crow's arguments—and in many cases directly quoted passages—found their way into the
articles and public lectures of many of genetic toxicology's most ardent promoters (Crow 1971a:24; Epstein 1969b; Legator 1970; Malling 1970; Meselson
1971). Its message resonated far beyond the mutation research community in
the form of congressional testimony (U.S. Senate 1971), newspaper articles and
columns (Lederberg 1969; Schmeck 1970a,b), articles in industry trade journals
(Sanders 1969a,b) and educational digests (Crow 1971b), and in at least one
book-length study (Turner 1970). Within just a few years, chemical mutagenicity became more than a key to unlocking the mystery of gene action; it also
became a measure of public health, an index of deleterious trends threatening
human evolution, and a call for scientific and political action. Today, "Chemical Risk" is considered a classic among first- and second-generation genetic toxicologists. It is a cornerstone of insider origin stories and continues to attract

attention as a foundational event in the historical development of genetic toxicology (Brusick 1990; Crow 1989; Prival and Dellarco 1989; Wassom 1989).

The retrospective attention that insiders have lavished upon "Chemical Risk," as well as the "shot heard around the world" impact the article had at the time of its publication, is somewhat paradoxical. "Chemical Risk" was not the first article to address the issues it raised, nor was the knowledge it imparted unfamiliar territory, especially for those working in mutation research. Nevertheless, the article struck a highly responsive chord among many in mutation research and beyond. Its publication on the eve of scientists' mobilization around the issue of "environmental" mutagenesis suggests that the ideas and concerns expressed in "Chemical Risk"—and perhaps even more important, the manner in which those ideas were packaged—played an unexpectedly important role.

The goal of this chapter is to show how scientist-activists made the rhetorical case for genetic toxicology. It offers an analysis of public lectures, expert testimony, articles, and editorials produced by several of the genetic toxicology movement's activist core during the 1968–1973 period. These promotional texts were not scientific papers; they did not present new data, describe new methods, or establish new facts. Instead, these texts functioned as "collective action frames." Collective action frames are "action-oriented sets of beliefs and meanings that inspire and legitimate the activities and campaigns of a social movement organization" (Benford and Snow 2000:614). The promotional texts presented here redefined synthetic chemicals in the human environment as *genetic* hazards, imbuing the biological problem of chemically induced mutations with new environmental and political meaning. The genetic hazard frame's rhetorical flexibility allowed scientist-activists to express grievances, attribute blame, and prescribe courses of collective action selectively as befit different contexts and audiences. Using Snow et al.'s (1986) concepts of "frame amplification" and "frame extension," I describe how scientist-activists recast chemical mutagens as an environmental health problem. In the process, they broadened the scope of genetics research, and the moral responsibility of geneticists, to include pollution and public health issues. Using a third frame alignment strategy that I call "frame translation," scientist-activists positioned mutation research as a key element in solutions to research problems in other fields (e.g., developmental abnormalities). Framed thus, chemical-induced mutations were not merely framed as relevant to the environmental health sciences but as a key to solving the problem of human exposure to synthetic chemicals. Frame translation helps to better account for the interdisciplinary nature of these mobilization efforts.

I am also interested in exploring briefly why these scientist-activists were invested in a frame that constructed chemical mutagenesis as environmental problems. A few of the core activists in this movement could reasonably be labeled moderate environmentalists.[1] But as a whole, the founding members of EMS who led the movement to establish genetic toxicology had far more enduring loyalties to human genetics and its promise to cure genetic disease than to toxicology and the emerging field of environmental health. They also harbored stronger commitments to the politics and social activism of an older, and largely discredited, eugenics movement than to a new wave of environmentalism. Joshua Lederberg, James Neel, and James Crow, in particular, were to varying degrees invested in rehabilitating a nonracist eugenics that put genetic knowledge, technology, and clinical practices to use toward the betterment of human society. Chemical mutagens threatened the long-term integrity of the human gene pool as well as the environmental health of living individuals and communities. The goals and interests of reform-minded eugenicists dovetailed with environmentalism in interesting ways in genetic toxicology. This chapter offers an interpretation of that convergence.

Constructing the Genetic Hazards Frame

"Chemical Risk" reads more like a political manifesto than a scientific tract: stylistically programmatic, notably bereft of explicit detail, and written for an educated lay audience, with a glossary replacing the typical list of references. It incorporated vivid imagery and a subtle but clear explication of what geneticists knew or did not know about environmental chemical mutagens to paint for readers a compelling diagnosis of the problem, recommendations for solving it, and, perhaps most important, a spirited call to action. "There is reason to fear," Crow wrote, "that some chemicals may constitute as important a risk as radiation, possibly a more serious one. Although knowledge of chemical mutagenesis in man is much less certain than that of radiation, a number of chemicals—some with widespread use—are known to induce genetic damage in some organisms. To consider only radiation hazards may be to ignore the submerged part of the iceberg" (1968:113). Crow pointed to the existence of certain chemical compounds that tested negative in standard toxicology screens and that failed to produce overt genetic effects (e.g., chromosome breakage) but that were nevertheless highly mutagenic in bacterial test systems. Geneticists thus faced a dangerously paradoxical situation in which chemical compounds that caused the least amount of genetic damage at an individual level were, in Crow's words, "most insidious" in terms of their potential long-term effects on the human population. "Even though the compounds may not be demonstrably mutagenic to man at the concentrations used," he warned, "the total number of deleterious mutations induced in the whole population over a prolonged period of time could nevertheless be substantial" (113).

The article spelled out two possible scenarios, both of which would portend potentially dire and irreversible consequences for the human species. In one, chronic exposure of large populations to low concentrations of mildly mutagenic compounds could induce damage to the gene material that, when passed down over several generations, might effect a slight average increase in the human mutation rate. This would result in a gradual and probably imperceptible decline in the genetic health of the population. Another scenario, which Crow labeled a "genetic emergency," involved a situation in which "some compound presumed to be innocuous is in fact highly mutagenic and that large numbers [of people] are exposed before the danger is realized" (1968:113–114).

To stem the (invisible) damage already done and to slow if not prevent further degradation of the human genetic material, Crow presented four recommendations for research and policy action. These included the creation of a chemical information database, the development of new bioassays for identifying chemical mutagens, routinized mutagenicity testing of new and existing chemical compounds, and the development and implementation of programs to monitor "at risk" human populations (Crow 1968:114, 116–117). Crow's recommendations for containing the invisible "iceberg" of chemical genetic damage suggested a technocratic approach that fixed social responsibility for organizing the rational management of environmental chemical pollutants and for preventing a "genetic emergency" on scientists, who would be empowered by the financial support and regulatory action of the federal government. As he presented the problem, "Chemical Risk" functioned as a rhetorical scaffold on which scientist-activists built the case for genetic toxicology.

Diagnostic, Prognostic, and Mobilizing Functions

In their promotional texts, interviews with reporters, and testimony before congressional hearings, scientist-activists borrowed the ideational elements contained in Crow's 1968 article. Importantly, they did not challenge the basic tenets of the various problems outlined in "Chemical Risk" but instead elaborated upon those basic ideas and concerns, filling in gaps and filling out the details of Crow's original summary. The following examples illustrate widespread and consistent trends.

First, the promotional texts provided a clearly articulated diagnosis of the manifestations of chemical pollution and its long-term implications, as in Sam Epstein's remark to a reporter for an industry trade journal: "At this moment we may be in the midst of a potentially serious accidental experiment on the effects of chemical mutagens in man—the full impact of which may not be known for generations to come" (Epstein, quoted in Sanders 1969a:50). Crow's (1968) critique of government intransigence in funding research on and regulating chemical mutagens was at best implicit. Scientist-activists, however, explicitly and repeatedly pointed to congressional inaction as a cause for concern. As Epstein told a Senate subcommittee hearing on "Chemicals and the Future of Man," "It is interesting to point out that within the last eighteen months three expert committees have unanimously recommended that mutagenic testing be made mandatory for food additives and pesticides. But as yet, no regulatory action has been taken" (U.S. Senate 1971:10).

Second, following Crow's lead, scientist-activists outlined a series of concrete prognostic steps for attacking the problem. An FDA geneticist warned the same Senate subcommittee, "We must begin to seriously monitor human population in ways to better evaluate the extent of the problem. We can continue to expand our testing of compounds, such as drugs and food additives for potential mutagenic activity and we must undertake more basic investigations predicated on our need to know how mutations occur and how they are repaired" (U.S. Senate 1971:29).

Perhaps most important, "Chemical Risk" provided the basis for a call for scientific collective action and increased public awareness. Future Nobel laureate Joshua Lederberg (1969) noted in one of his regular "Science and Man" columns that "we [biologists] all have a basic responsibility to go beyond an emotional expression of concern; to use it to energize the search for authentic scientific measures of potential hazards and for means to neutralize them. . . . We have a great deal of taxing work ahead in trying to set up scientifically valid and politically useful criteria from laboratory studies for these elusive but all-important hazards."

Over time, the basic structure and content of these integrated arguments did not change in their elemental form. In this case, the frame's flexibility paradoxically conditioned its stability. Scientist-activists interpreted the dangers noted in "Chemical Risk" differently in different contexts, all the while remaining allied with Crow in their generalized expressions of concern.

Rhetorical Flexibility

Collective action frames tend to be loose and relatively informal sets of ideas, rather than formal ideological systems, because they need to be flexible in order to adapt to changing situations (Tarrow 1992:190). "Chemical Risk" provided a broad range of concerns from which scientist-activists could draw selectively and elaborate in papers and public presentations before policy makers, their research peers, university administrators, and high school biology classes. Accordingly, one could frame the problem of environmental chemical mutagens with equal flair as an economic burden, a moral dilemma, or a natural disaster.

At the NAS symposium "Aids and Threats to Society from Technology," James Neel emphasized an economic cost-benefit calculus in assessing chemical technology:

> As we struggle to move towards those new levels of social and techno-
> logical organization which will enable us to meet the kinds of problems
> we have been discussing in this symposium, surely an important ele-
> ment in the decision-making process must be knowledge of the genetic
> cost. . . . [I]n order to support our technology we may have to compro-
> mise with the desire of the geneticist for no increase in mutation rates—
> but we owe it to our offspring to see that the compromise is based on
> knowledge rather than a guess we may later regret. (Neel 1970:913–914)

But in a different context—the inaugural celebration of a toxicology laboratory
in Germany—Sam Epstein came to quite different conclusions using the same
basic imagery and logic:

> It has recently been estimated that in the U.S.A. the total costs to society
> of one malformed child are in the region of one million dollars. Such
> estimates can also be made for chronic toxicity and for carcinogenicity,
> as the immediate impact of these hazards are restricted to one individ-
> ual. However, for genetic hazards, we cannot predict the future extent
> and degree of damage to succeeding generations. The costs may well be
> incalculably high. For these reasons, the concept of matching benefits,
> which we have generally accepted in toxicology, are probably quite inap-
> propriate for genetic hazards. (Epstein 1969c:9)

Crow's (1971a) concurring statement illustrates the moral dilemma at stake:
"There is obviously no simple and realistic system of cost-accounting. (How
many gastric ulcers equal one childhood death?)."

The nature of the genetic hazards also could be interpreted flexibly. If one's
audience were local citizens, one might choose to emphasize the visible, near-
term effects of genetic damage. Oak Ridge geneticist Fred de Serres had this to
say to some of his fellow Tennesseans: "[W]e already have a large number of
individuals suffering from genetic diseases in our society. Genetic defectives are
not only a personal tragedy if they happen to occur in your own family, but they
may wind up as a tremendous burden to the population in general, since they
require expensive medical care and special institutions."[2] In front of an audience of biologists, alternatively, emphasis could be placed on the evolutionary
long view, as in Marvin Legator's essay published in the *Journal of Heredity*:
"While lethal and sub-lethal mutations will be rapidly eliminated from the population, and need not be of great concern, nonlethal mutations will tend to per-
sist through several generations with the duration of their persistence being
inversely proportional to the severity of their effects" (1970:253).

The relationship between knowledge and action was another fulcrum
around which the promotional texts shifted. The existing state of knowledge
could be cast as a resource or as a liability, as it was seen to fit a particular

context. If one's audience was the mutation research community, it made some sense to focus on existing knowledge as a basis for action. Thus did Harvard biologist Matthew Meselson caution that

> just as in the case of carcinogenicity testing, there will be important differences in response between different systems for mutagenicity. However, this is no excuse for doing nothing. . . . Indeed, as compared with cancer testing, we are already in a better position to conduct meaningful tests since we have a fairly good understanding of the molecular basis of mutation but no comparable insight into carcinogenesis. (Meselson 1971:xi)

Alternatively, if one's goal was to attract biologists from other subfields into mutation research, emphasizing the sizable gaps in existing knowledge might prove a more effective rhetorical strategy. In an address to radiation biologists, for example, Alexander Hollaender remarked, "Then, of course, there is the problem of mutagenesis. . . . Here were have very little hard data in regard to mutagenesis produced by chemicals. As a matter of fact, we know as little about this as we knew about radiation in 1930."[3]

A flexible frame allowed scientist-activists to tailor their messages to particular audiences. Moreover, the language used in these promotional texts underscores the importance of the particular contexts in which scientist-activists found themselves, be they congressional hearings, scientific conferences, or high school classrooms. Frame analysis reminds us that forging an effective link between arguments and practices constituting an emergent genetic toxicology, on the one hand, and diverse research communities and patrons, on the other, was never a foregone conclusion; the modifications that scientist-activists made by framing chemical mutagens as genetic hazards reflected perceptions about what arguments were likely to mobilize different audiences most efficiently. This analysis of the genetic hazards frame's core framing functions helps us understand what scientist-activists themselves understood as pockets of boundary permeability and which groups they identified as cadres of potential support. I turn next to an analysis of the three "strategic framing processes" (Snow et al. 1986) that scientist-activists used to link their case for building genetic toxicology to the interests, values, and beliefs of potential recruits and supporters.

Frame Amplification

Frame amplification involves "the clarification and invigoration of an interpretive frame that bears on a particular issue, problem, or set of events" (Snow et al. 1986:469). Scientist-activists amplified the genetic hazards frame in an attempt to convince biologists already familiar with chemical mutagenesis as a research tool that mutagenic chemicals also constituted a serious environmental problem. The connection, for many research geneticists, was not obvious or automatic.[4] Scientist-activists developed three specific discursive techniques for amplifying the relationships between chemical mutagens in the laboratory and the health implications of chemical pollution.

The first involved underscoring the environmental and health implications of existing knowledge in mutation research. A preface to a collection of studies describing methods of mutagenicity testing noted that "our knowledge of genetics and molecular biology clearly establishes the possibility that exposure of human germ tissue to certain exogenous agents can cause genetic damage" (Meselson 1971:ix). Another text reminded geneticists of "the cumulative importance of mild effects" by providing a basic primer on five generalizations borne out by mutation research on fruit flies (Crow 1971a).[5] Here, "the facts" of mutation research were presented in such a way as to make the human health implications of exposure to chemical mutagens appear obvious (and obviously detrimental).

Another frame amplification technique involved emphasizing the ubiquity of chemical pollution. For example, Hollaender used the occasion of a symposium on environmental pollutants to remark to his audience of radiation biologists that the "list of pollutants is long and the paucity of information on their long-term effects is staggering."[6] Others were considerably more specific, providing lists of industrial processes such as printing, dyeing, and fireproofing (Epstein 1969a), as well as products:

> These new man-made chemicals are everywhere. Some 2–3000 are used as food additives; 30 are used as preservatives; 28 as antioxidants; 44 as sequesterants; 85 as surfactants; 31 as stabilizers; 24 as bleaches; 60 as buffers, acids or alkalies; 35 as coloring agents; 9 as special sweeteners; 116 as nutrient supplements; 1077 as flavoring agents, and 158 for miscellaneous uses. Thousands of other compounds are used as drugs, narcotics, antibiotics, cosmetics, contraceptives, pesticides or as industrial chemicals.[7]

Lists like this one represented a stark departure from the way that chemical mutagens typically were represented in scientific journals as tools of analysis.

The third amplification technique involved distinguishing chemical mutagenicity from other physiological processes. It was not uncommon in these promotional texts for scientist-activists to point out the levels of complexity that chemical mutagens present for the basic researcher, relative, for example, to radiation. In one text, Hollaender noted that "there are only three or four types of radiation, but there are produced each year about 30,000 new compounds, of which probably 20 or 30 are mutagenic."[8] Other texts emphasized the physiological differences between mutagenicity and toxicity: "We must also remember that, by its nature, this genetic damage can be cumulative over generations while even the most insidious nongenetic poison cannot accumulate in the body beyond the lifetime of an individual" (Meselson 1971:ix).

In each of these examples of frame amplification, the information being stressed was not new knowledge. Amplifying the genetic hazards frame mainly involved highlighting relatively well established understandings of genetic processes and chemical-biological relationships already circulating in the pool of common knowledge in genetics. But as Snow and Benford (1992:138) have noted, "What gives a collective action frame its novelty is not so much its innovative ideational elements as the manner in which activists articulate or tie them together." In this case, the innovation comes in demonstrating through persuasive rhetoric that chemical mutagens had "lives" outside the laboratory.

Frame Extension

Another framing process, called "frame extension," involves "extending the boundaries of [the] primary framework so as to encompass interests or points of view that are incidental to its primary objectives but of considerable salience to potential adherents" (Snow et al. 1986:472). Scientist-activists extended the genetic hazards frame in two basic ways. One involved extending the scope of chemical mutagenesis itself beyond genetics. The other involved extending the scope of solutions to include critical analyses of the political, organizational, and disciplinary structures impeding the pursuit of genetic toxicology research. Both strategies increased the mobilization potential of the movement by casting a wider rhetorical net around the problems and solutions implicated by the genetic hazards frame.

By far the most prevalent technique for extending the scope of the chemical mutagenesis problem beyond a strict genetic interpretation was accomplished by casting chemical mutagenesis as a central and critical dimension of environmental pollution. In a paper titled "Genetic Implications of Pollutants," for example, Hollaender observed that "of all the environmental pollutants to which man is exposed, the greatest uncertainty concerning disastrous effects surrounds those which have an effect on man's genes and chromosomes — the carriers of inheritance. . . . The field of chemical mutagenesis is developing rapidly, and is one of the most important factors in environmental pollution."[9] An article by British geneticist Bryn Bridges (1971:13), originally published in the *Ecologist* and later reprinted in the *EMS Newsletter*, begins, "Environmental pollution and its attendant hazards are newsworthy, and rightly so. The risk of genetic damage to man is one aspect, however, which has so far received less attention than it deserves, at least here in the U.K."

Another strategy for extending the genetic hazards frame involved a broadranging critique of the social structure of environmental research. In general, scientist-activists framed chemical genetic hazards as the direct outcome of regulatory inaction, noting that no government regulations required mutagenicity testing for new or existing chemicals.[10] Additionally, however, some also extended their critique to include a number of underlying social-structural and ideological factors.

In the private sector, the economic interests of chemical and pharmaceutical industries, for example, often "led to somewhat restricted approaches to toxicological problems—narrow questions, narrowly defined, narrowly posed and often narrowly answered."[11] Direct clients of industry, commercial testing laboratories suffered similar constraints and generally failed to give the problem of mutagenicity testing the attention some thought it deserved (Hollaender 1973:232). Further, by restricting outside access to their data, both industry and commercial testing laboratories blocked regulatory agencies and policy decisions.

In regulatory agencies, the compartmentalization of research on chemical effects into separate agencies for dealing with the environmental, consumer, and occupational dimensions of environmental/health problems hampered the development of programs to train people who were not engaged in basic mutation research to conduct mutagenicity tests competently, since "each government agency [thinks] someone else should do it" (Hollaender 1973:232). Environmental research at the national laboratories, ideal places for intensive research programs, was similarly constrained by "conflicting regulatory, promotional, and research roles."[12] At universities, traditional departmental structure impeded interdisciplinary interaction and the efficient and creative use of their uniquely concentrated resources (Lederberg 1969).

Finally, scientist-activists argued that institutionalized differences between "basic" and "applied" science placed limits on environmental research at all of these institutions, resulting in a piecemeal approach to problem-centered research (Lederberg 1971; Meselson 1971). In combination, scientist-activists argued, these structural constraints and their reinforcing ideological divisions inhibited the free flow of knowledge, data, people, and practices across existing institutional and also conceptual divides. Promotional texts framed this "artificial fragmentation" as antithetical to the orchestration of systematic, multidisciplinary attacks on complex problems in the environmental, occupational, and consumer health sciences.[13]

Frame Translation

It is important to underscore the depth of the institutional obstacles facing these scientist-activists. As noted in the previous section, their promotional texts cited numerous barriers to interdisciplinary collaboration. The scientific status of genetics added further to the problem. While fast gaining prominence in American life science, genetics circa 1970 was far from the powerful discipline it is today. The biotechnologies that have made genetics relevant to nearly all life science fields today were only beginning to be developed (Kenney 1986); cancer was not yet understood as a "genetic disease" (Fujimura 1996); and Ph.D. requirements in toxicology did not typically include genetics courses.[14] In effect, the social structure of research impeded scientist-activists' attempts to confront in chemical mutagenesis "a fantastically complex problem—one that, in many ways, is nebulous and hideously difficult to come to grips with" (Crow, quoted in Sanders 1969a:71). Establishing the EMS was one way that scientist-activists sought to "provide a channel for communication among a wide range of separate disciplines" (Lederberg 1969). Another related approach for enlisting researchers into the EMS and into genetic toxicology was frame translation.

I borrow the term "frame translation" from Bruno Latour's (1987:117) notion of "translating interests" as he uses it to describe processes of fact-making in technoscience. For Latour, "'interests' are what lie *in between* actors and their goals, thus creating a tension that will make actors select only what, in their own eyes, helps them reach these goals amongst many possibilities" (108–109). Translation is a strategy for gaining position with respect to one's colleagues and competitors by reinterpreting others' interests in terms of one's own. Boiled down to its essentials, interest translation involves skillful marketing campaigns. In Latour's classic illustration, in order to "pasteurize" France, Louis Pasteur had to convince farmers, veterinarians, and eventually the rest of French society that their interest in controlling anthrax and other diseases would be best served by believing in him, in his vaccine, and in the microbes that he had isolated and made visible. To keep French cattle alive, in other words, farmers had to recognize that what Pasteur controlled in his laboratory was indispensable to them (Latour 1988).

Similarly, frame translation describes a process in which potential supporters' own specific interests are best served by adopting the same goals or employing the same practices as movement activists. In genetic toxicology, the rhetorical influence of frame translation lay in connecting disparate points of interest and in creating "detours" that made environmental mutagenesis necessary to other scientists' projects (Latour 1987:116). This process worked in multiple directions as a strategy for enrolling geneticists in efforts to educate public health scientists and the regulatory community and as a strategy for enrolling toxicologists and other health scientists not familiar with mutation research practices. More than frame alignment and frame extension, frame translation illustrates how a small group of committed geneticists worked to consolidate genetic toxicology as a broadly interdisciplinary science with genetic theories of mutation at its core.

The most committed scientist-activists promoted their concerns "abroad" in forums devoted to medicine, toxicology, cancer research, radiation research, and reproductive biology and in the NAS. In this subset of documents, frame translation processes are particularly evident in scientist-activists' efforts to elaborate the interrelationships between mutations (mutagenesis), cancer (carcinogenesis), and developmental abnormalities (teratogenesis). An essay appearing in a medical science newsletter argued that "the major effects of increased mutation rates would be spread insidiously over many generations, and would include ill-defined abnormalities, such as premature aging, alterations in sex ratios, and increased susceptibility to various diseases, including leukemia and cancer" (Epstein 1969a:11). At a conference on "Methods for Detecting Teratogenic Agents," Hollaender used the speaking time allotted him as moderator of a panel session to talk about the problem of chemical mutagenesis and reviewed three popular methods for the detection of chemical mutagens.[15] The most pointed illustration of frame translation I found comes from an essay, "Chemical Mutagenesis Comes of Age":

> Aflatoxin [a fungus] is teratogenic, carcinogenic and mutagenic; cyclamate [a synthetic sweetener] has been shown to induce bladder tumors in addition to its cytogenetic effects; Captan [an herbicide] was found to be teratogenic in laboratory animals and DDT produces tumors in animals. Although it need not follow that a chemical producing one response necessarily produces the others ... there are similarities among agents that produce hereditary alternations in the information content or distribution of hereditary material whether the final expression is mutagenic, teratogenic, or carcinogenic. (Legator 1970:255)

In this passage, Legator is very clearly crafting a "detour" as Latour uses the term: scientists engaged in problems of tumor induction or birth defects are best served by paying attention to the work being done by mutation geneticists. Chemical mutagenesis is a research problem that can provide answers to problems of tumor induction and birth defects. The causes of both are linked to gene mutation and also to environmental chemical pollution.

Frame translation also served a specific mobilizing function that encouraged cross-disciplinary interaction in laboratory and regulatory practice. The same essay implores readers to break the shackles of insular disciplinary cultures:

> For far too long, toxicologists have concerned themselves with genetic problems without the needed expertise of geneticists. The problem of selection of hybrid or inbred strains of experimental animals, the interpretation of results from reproductive, carcinogenicity, or teratology studies surely requires the expert knowledge of both toxicologists and geneticists, and their cooperative participation is long overdue. (Legator 1970:255–256)

A similar dynamic appears in this suggestion that mammalian mutagenicity testing be made part of the standard repertoire of toxicology screens: "[M]utagenicity tests must be integrated into routine toxicologic practice, even to the possible future extent of parallel investigations in toxicity, carcinogenicity and teratogenicity in certain toxicological institutes" (Epstein 1969c:10).

In these and other examples, scientist-activists engaged in frame translation. Their published and spoken commentary promoted knowledge about chemical mutagenesis and the technologies that geneticists had developed for identifying chemical mutagens as enabling the growth of knowledge in other biomedical and public health fields. They also promoted an understanding of chemical mutagens as an environmental problem and so encouraged research and regulatory activism from a wide range of specialists. Genetic toxicology was explicitly defined in these promotional texts as a broad-scale multidisciplinary effort to understand, identify, and prevent the long-term genetic deterioration of human populations. It was framed as a social solution to the institutional problems—disciplinary, economic, and cultural—that beset the health sciences and as a scientific solution to the environmental problem of chemical pollution.

As some science studies scholars have noted, translation works in multiple directions; those doing the translation are also often at the same time being translated by others (Fujimura 1992; Star and Griesemer 1989). Similarly, frame translation served an important function in the context of the movement to establish genetic toxicology by casting doubt on the efficacy of taken-for-granted disciplinary and other social boundaries separating genetics from public health sciences. As a rhetorical strategy engaged to build an interdiscipline, frame translation attracted outsiders in, but it also urged insiders to step out into less familiar territory. This dialogical view also underscores the point that frame alignment is a dynamic interactive process such that "there is no sharp demarcation possible between beliefs that are 'within the system' and those that are 'outside the system'" (Tarrow 1992:190). The rest of this chapter builds on Tarrow's insight by showing how the genetic hazards frame linked discursively to the broader environmentalist culture from which scientist-activists drew many of their most culturally salient images, language, and ideas.

Frame Resonance and Environmentalist Political Culture

To attract new adherents and political allies, social movement actors must construct collective action frames that resonate with the audiences they're attempting to mobilize (Snow and Benford 1988). For this very practical reason, most collective action frames are knit from existing cultural resources—the values, beliefs, and ideas that are embedded in cultural tradition. Collective action frames are not, in other words, cut primarily from new cloth. Rather, they are stitched together from the old and familiar. However, frames constructed from the existing popular culture may reinforce complacency rather than inspire action to the extent that they embody and reflect normative social arrangements or status quo sensibilities. In order to provide the emotional or ideological inspiration that is often necessary to spur people to engage in social protest, collective action frames also must incorporate some elements of what Tarrow calls an "oppositional political culture"—the value sets that are "rejected by many but are inherited from the society's tradition of collective action and opposition" (1992:192). Collective action frames thus embody a relational process, the weaving together of values, beliefs, and meanings that are both old and new, mainstream and avant-garde, dominant and subversive.

The genetic hazards frame seems to have enjoyed a relatively high degree of resonance within the biological and public health sciences communities targeted by genetic toxicology scientist-activists. It is also fairly clear, judging from the above examples, that the burgeoning environmental movement provided a "reservoir of symbols" from which movement organizers drew many of the frame's ideas and images (Tarrow 1992:197). This claim in itself is not new. The existence of a connection between the motivations of scientist-activists and "environmentalism" writ large is taken for granted in insider accounts of the rise of genetic toxicology (e.g., Prival and Dellarco 1989). In order to understand why the genetic hazards frame was so effective at arousing interest and inspiring action, however, it is useful to consider in greater detail the factors underlying and contributing to this frame's relative success.

Snow and Benford (1988:207–211) again provide the analytic tools for carrying out this task. They have proposed three sets of "phenomenological" factors that shape a frame's potential to mobilize constituents. According to their argument, frames need to be seen by target audiences as having "empirical credibility": can the claims posited by the frame be verified? Frames also need to project a level of "experiential commensurability": does the frame harmonize with people's lived experiences? Finally, frames require "narrative fidelity": does the frame "ring true" with existing popular understandings? We can apply these analytical categories to the genetic hazards frame, doing so in reverse order.

Narrative Fidelity

Rachel Carson's *Silent Spring* provided the main narrative for the new environmentalism of the 1960s and the 1970s (Brulle 2000), and her view of chemical mutagens is instructive. "For mankind as a whole," Carson wrote, "a possession infinitely more valuable than our individual life is our genetic heritage, our link with past and future. Shaped through long eons of evolution, our genes not only make us what we are, but hold in their minute beings the future—be it one of promise or threat. Yet genetic deterioration through man-made agents is the menace of our time, 'the last and greatest danger to our civilization'" (1962:186).

Silent Spring is a tale of the unintended destruction of natural systems as the threatened result of corporate nearsightedness and the sometimes willful ignorance of policy makers and scientists. This storyline also provided much of the narrative basis for the genetic toxicology movement.[16] "In the past," wrote James Crow, "we have been quite reckless in our ignorance. New chemicals have been widely used long before much was known about their long-term effects on either man or wildlife" (1971a:105). Pursuing a similar theme, de Serres noted in a speech to concerned citizens, "What supreme arrogance and folly it would be for us to imagine that we are immune to the dramatic and deadly consequences of environmental chemical pollution and to sit back and do nothing about such a serious problem! If we do nothing, we may well find ourselves in the position of the California Brown Pelican—just another animal species on the road to extinction!"[17] And using a rhetorical style that meshes almost seamlessly with Carson's (1962:14) dark specter of "a spring without voices," Gary Flamm told a Senate subcommittee that "in the final analysis our genes are our most important legacy—all else is secondary by comparison—for if our gene-pools become overloaded and overburdened by deleterious mutations, our future as a species is indeed bleak" (U.S. Senate 1971:29).

Also like Carson, the genetic toxicology activists who further elaborated the imperatives set out in Crow's 1968 article drew rhetorical sustenance from much older preservationist and conservationist discourses (Brulle 2000). The preservationist impulse to honor and protect wilderness, in particular, is widely apparent in the rhetorical construction of human genes as scarce and fragile natural resources. In their public addresses and writings, geneticists repeatedly made references to the uniqueness of the human genetic material using language evocative of the mid-nineteenth-century romantic ecology of Henry David Thoreau or the early-twentieth-century pastoral spiritualism of John Muir (Worster 1994:58, 16–17). "What the public must recognize," James Neel told an interviewer, "is that mankind's most vital asset is not its material wealth but its germ plasm—the very stuff of life. Since the germinal cells are what determine the health, intellectual capacity, and all the other prime attributes of future generations, everything possible must be done to protect those—humanity's most precious possessions" (quoted in Sanders 1969a:71). The stakes in the roulette game that society was unwittingly playing with chemical mutagens were such that, as Joshua Lederberg put it, "no one can be totally indifferent to his responsibility as a vessel of the species, to a role in human evolution that answers to the most profound religious instincts" (1969). Marvin Legator reasoned that "the protection of man's genetic heritage is one of the most critical issues of our times," and Gary Flamm warned that "many scientists, particularly geneticists, are deeply worried whether man's stewardship over his own genetic material might not be tragically inadequate" (U.S. Senate 1971:27).[18]

If genetic toxicology activists took advantage of Carson's dire predictions of ecological deterioration that she used to introduce readers to the argument of *Silent Spring* as a way of rousing complacent scientists and policy makers, they likewise adopted the portentous tone with which she ended her book. "We stand now where two roads diverge," Carson observed in the opening paragraph of *Silent Spring*'s final chapter: "But unlike the road in Robert Frost's familiar poem, they are not equally fair. The road we have long been traveling is deceptively easy, a smooth superhighway on which we progress with great speed, but at its end lies disaster. The other fork of the road—the one 'less traveled by'— offers our last, our only chance to reach a destination that assures the preservation of our earth" (1962:244). There is a strong sense of immediacy in the promotional texts I examined, suggesting that geneticists adopted imagery similar to that in *Silent Spring* in order to create a climate of urgency and a shared sense of social responsibility to act collectively *as scientists* to prevent environmental chemicals' unintended genetic consequences. This is illustrated by the statements of a number of geneticists who argued that the lack of direct data pertaining to chemical mutagenicity in humans is "no excuse for doing nothing" (Meselson 1971:xi). "If we do have a serious problem," urged Frederick de Serres in a statement that was uncommon only in the intensity of its symbolism, "research programs must be started immediately to determine how best to preserve the genetic heritage that makes man unique, among all of the animal species that inhabit this planet."[19] And at a presentation before members of the American Chemical Society, NIH molecular biologist Ernst Freese pressed the issue. "We should not wait for decades of statistical evaluation," he argued, "but start now to reduce our exposure to potential mutagens—at least until they have been proved harmless. If the possibility of cancer does not seem threatening enough to everyone, at least young people owe this precaution to their future children, who should not be doomed to lives devastated by genetic defects brought about by chemically induced mutations" (quoted in Sanders 1969a:52).

Importantly, this sense of impending and irreversible crisis imbued the genetic hazards frame with a rhetorical energy that was itself derived from the same evolutionary perspective that drove Carson's arguments in *Silent Spring*. The book became a foundational text for the new environmentalism of the 1970s because it framed the problem of the environment largely in terms of long-term, indirect, largely invisible, and unintended effects of chemical pollution.[20] Analogously, the genetic hazards frame fashioned by genetic toxicology activists less than a decade later was not concerned primarily with the genetic diseases of the current population—with somatic mutations that manifest in tumors or with gross chromosomal aberrations that manifest in developmental abnormalities such as Down's syndrome. Instead, these geneticists were concerned with the effects of minute recessive mutations that could lie dormant for generations. As James Neel and Arthur Bloom (1969:1254) reminded their readers, "If we introduce mutagens into the environment, it is not so much ourselves as subsequent generations who pay the price." This is an important point, which we will revisit in the following chapter as well: the portending crisis was not one of this generation but of seven or more generations hence. Genetic integrity of future populations was the central theme of the genetic hazards frame. It was one that "resonated" with other scientists, and it was a message also that, because its narrative structure was by 1969 largely a familiar one, was likely to resonate clearly with the environmental movement's popular base.

Experiential Commensurability

In contrast, the level of experiential commensurability of the genetic hazards frame among the general public, as among those audiences it specifically sought to mobilize, was probably relatively low—this for the simple reason that most people did not (and do not) consciously experience the effects of germ-cell chemical mutagenesis. Survey research data from the period does suggest, however, that public awareness of environmental problems more generally was increasing (Dunlap 1992:91–96). In the years following the 1962 publication of *Silent Spring*, fish kills, the disappearance of wildlife, ecological damage to lakes and streams, acid rain, brown clouds, and outbreaks of mysterious livestock and human illnesses in countries around the world belied a widening "circle of evidence" reported in the popular and scientific press that lent support and urgency to Carson's claims (Graham 1970:95; Cairns 1968; Henry 1971; Wit et al. 1970; for a general survey of environmental ills, see Hayes 1987). The genetic effects of chemicals in the environment further expanded the circle. The FDA's decision in 1969 to ban the artificial sweetener cyclamate was one of the more highly publicized actions taken to restrict the use of known mutagens (Epstein et al. 1969; Turner 1970). There also appeared newspaper and trade journal articles covering topics such as the mutagenicity of gasoline additives and pharmaceutical and recreational drugs (Sanders 1969b; Schmeck 1970b). Another report noted advances in new methodologies for monitoring human populations for increases in the rate of mutation (Schmeck 1970a). The media attention given mutagenic chemicals during this period meant that the general public had at least some occasional indirect knowledge of their deleterious effects.

In the scientific community, and specifically among research biologists, the level of experiential commensurability with environmental mutagens can be assumed to have been somewhat higher in the sense that the vast majority of the plants, insects, microorganisms, and animals that populated biology laboratories were the living results of chemical or radiation mutagenesis. And for at least the few geneticists who ventured into the natural environment for field research, their experiences with "environmental" mutagenesis were still more direct. Geneticists at the Oak Ridge National Laboratory, for example, obtained funds from the National Science Foundation Environmental Studies Program to study the effects of mercury contamination in the nearby Boone Reservoir (Malling et al. 1970). For others, environmental contaminants came (uninvited) to them. At Brookhaven National Laboratory, the accidental release of mutagenic chemicals into a greenhouse filled with *Tradescantia* plants provided geneticists there with the opportunity to observe a "natural" experiment (Sparrow and Schairer 1971).[21]

For most people, scientists and nonscientists alike, the commensurability of the genetic hazards frame might be described more accurately as "vicarious" rather than "experiential." In the main, environmental chemical mutagenesis was lived through the scientists who produced the studies and was filtered once again through the media. It is thus telling that genetic toxicology activists seem to have made some effort to bring their audiences closer to their mutation research laboratories by using imagery in their promotional texts that readers could relate to as both familiar and tragic. They did so by drawing analogies between environmentally induced mutation in humans and various other well-known environmental catastrophes. Three events stand out: the increasing rate of birth defects, the thalidomide tragedy of the early 1960s, and the various fish kills disrupting lake and river ecology in the United States and Europe (e.g., Legator 1970).

What is interesting about all of these examples is that, while a clear rhetorical connection is being drawn between these publicly salient catastrophes and the essentially invisible problem of environmental mutagenesis, none of the former could be specifically and directly attributed to deleterious mutations. Rising rates of birth defects say little about their specific causes, which might arise from gene mutations but also might not. Thalidomide, for example, turns out not to be mutagenic at all, operating instead as a teratogen to disrupt fetal development (Sanders 1969a:54). And while mercury and the industrial and agricultural chemicals responsible for fish kills were indeed found to be mutagenic, it was their acute toxicity *as poisons* and not the disruption of DNA base pairs that resulted in the sudden death of fish populations. Nevertheless, these comparisons facilitated the link between the visible manifestations of environmental contamination and the largely invisible and long-term dangers of germ-cell chemical mutagenesis. That a level of commensurability with the genetic hazards frame existed at all says less about people's direct phenomenological experiences with genetic disease and more about genetic toxicology activists' efforts to construct a vicarious connection between existing and potential environmental tragedies.

Empirical Credibility

The promotional texts make clear that while relatively few chemical compounds had actually been tested for mutagenicity, enough data had accumulated to suggest that the potential for grave genetic hazard existed. Indeed, one of the main missions of these texts was to summarize and interpret the available evidence.[22] Based on the data at hand, four conclusions were routinely drawn:

1. Several workable mutagenicity bioassays existed. These were imperfect, to be sure, but improvements were under way, and new bioassay systems were in development.
2. Using these systems, geneticists had identified several classes of chemicals that were mutagenic at given concentrations.
3. Many of the chemicals found to be mutagenic in laboratory tests were used in agriculture and industrial manufacturing and were present in consumer goods and pharmaceutical drugs.
4. Chronic exposure to a wide array of chemical mutagens by large numbers of people was nearly certain.

These claims seem to have engendered considerable authority; I found no evidence that geneticists or any others questioned their basic validity.

Other claims that scientist-activists advanced as part of the genetic hazards frame, however, were considerably more susceptible to challenge. One issue that weakened the frame's empirical credibility involved questions concerning the load of human mutations. Was the rate of mutations in the human population actually increasing? Many scientist-activists suggested that it was. An increasingly mutagenic environment, they argued, in combination with the slowing down of natural selection through medical advances (that enabled people born with genetic disease to survive to reproductive maturity), virtually ensured an increase in the mutational load. This argument appears often in the promotional texts I examined, and the consistency of its rhetorical logic is impressive (e.g., Crow 1971b; Lederberg 1970; Legator 1970; Neel and Bloom 1969). However, scientist-activists could offer no empirical evidence confirming their theory. The necessary data did not exist for anyone to know with any degree of certainty whether the human mutation rate was in fact on the rise. And even if the data were to become available, "the problem of pinpointing the responsible agent," Neel and his colleague Arthur Bloom pointed out, would remain "formidable" (1969:156).

Another area of ambiguity in the genetic hazards frame involved the relationship between mutagenicity test results and human genetic health. "As far as chemical mutagenesis in man is concerned, we can at present only conjecture by analogy from results obtained in microorganisms, plants, animals, and cell cultures" (Ernst Freese, quoted in Sanders 1969a:52). Were the results of laboratory tests meaningful with respect to humans? What, if anything, did mutagenicity data from a bacteria bioassay tell scientists about genetic risk in people? Here, too, scientist-activists lacked the data needed to answer these questions with confidence. At the center of this debate was the issue of how far one might push inferences about the effects of genetic insults in one organism based on research conducted in another. I call this the "problem of extrapolation." It was in some ways even more intractable than the problem of measuring mutation rates.

The technical dimension of the extrapolation problem hinged on mutagen specificity. As we saw in Chapter 2, mutagen specificity refers to the fact that a single mutagen can cause "an almost infinite range of effects" depending on the organism, the method of delivery, the length of exposure, the concentration of dose, metabolic functions, and any number of other factors (Sanders 1969a:62). In an earlier context, the specificity of chemical mutagens was touted by some research geneticists as preferable to radiation as a research tool because of the greatly enhanced variations that could be achieved and the opportunities they offered for comparative research. In the more politicized context of the genetic toxicology movement, mutagen specificity presented a major, multidimensional stumbling block for scientist-activists promoting policy-relevant research on environmental genetic hazards. There were two underlying dimensions to the problem, one economic and one moral.

The infrastructure of mutation research provides an economic backdrop for understanding the problem of extrapolation and the dilemma it posed to the empirical credibility of the genetic hazards frame. The sunk costs of mutation research in terms of skill sets, procedures and experimental protocols, technologies, and experimental design all favored continuing to add incrementally to the knowledge base using existing resources. Chemical mutagenicity in humans was not a major component of the mutation research infrastructure, which instead was organized for the study of nonhuman standardized organisms such as microorganisms, plants, insects, and mice. From an economic standpoint, making small adjustments in existing research practices served geneticists' professional interests far more than the prospect of instigating the major changes to their established research programs it would have taken to retool their laboratories and acquire new skill sets for studying germinal human mutations.

More important, however, were the moral implications of in vivo mutagenicity testing on human subjects. As well as being highly impractical, geneticists considered exposing living humans to potentially dangerous chemicals for experimental purposes morally unacceptable (Sanders 1969a:56).[23] Population geneticists interested in the question had for some time relied on epidemiological methods to study human groups that shared some interesting genetic trait (Neel 1970) or who had experienced some specific mutagenic insult, such as the atomic bomb survivors at Hiroshima and Nagasaki (Neel and Schull 1956). These indirect methods of collecting evidence on human germinal mutation were quite limited in the level of precision they afforded as well as requiring "an extensive surveillance network" (Neel and Bloom 1969:156). Mutagenicity tests conducted in vitro provided another option that could be used as surrogates for in vivo tests. But these, too, were hampered by the problem of extrapolation. How can one be sure that a chemical testing positive in a bioassay of somatic human white blood cells, for example, will act the same way on human germ cells? Do tests of chromosome breakage reflect an accurate measure of the risk of point mutations (Sanders 1969a:58–59)? In sum, anything shy of in vivo tests in humans was susceptible to the charge of overinterpretation of the available experimental data. These vulnerabilities set up an interesting puzzle.

A Green Eugenics?

The correspondence of the genetic hazards frame with the symbols and interests of the nascent environmental movement is an important part of the explanation for the frame's resonance and subsequent success at mobilizing a critical mass of concerned scientists, administrators, and policy makers. But as we have just seen, the frame had its weak spots, particularly with respect to the authority of some of its more central empirical claims. The mobilizing power of environmental symbolism does not in itself sufficiently explain the ideological persuasiveness of the genetic hazards frame, particularly with respect to those potential adherents who may have mattered most: other research geneticists.

Moreover, as shown in the previous chapter, the core activists of the genetic toxicology movement—many of whom authored the promotional texts examined in this chapter—were seasoned scientists with solid reputations and records of research; they were not upstart environmentalists. With the minor exception of a few oblique references to student demands for university responsibility and the social and environmental implications of science and technology, there is little evidence that scientist-activists publicly aligned themselves with environmentalist or other "counterculture" movements.[24] Nor does the historical record suggest that most of these scientist-activists were generally committed to the larger political aims of the environmental movement per se. Rather, for most, their activism *as scientists* was highly specific to environmental chemical mutagens. In other words, deeply held individual commitments to environmental activism also do not explain why the genetic hazards frame seems so successfully to have fanned the flames of public interest science among established biologists.

The frame's subtle rhetorical symbiosis with the language of eugenics, however, may provide an additional important clue. There is evidence to suggest that the genetic hazards frame succeeded in part because it effectively dressed up old eugenicist fears of genetic degradation in the symbolic mantle of environmentalism, giving political vitality to a tired and suspect cause.

The connections linking the eugenics movement to both leftist and conservative geneticists in the United States are well established (Kevles 1985; Paul 1984, 1995). In the aftermath of the Nazi atrocities committed during the Second World War, the eugenics movement in the United States fell into widespread disrepute, with most geneticists distancing themselves from the racism the movement had come to symbolize (Paul 1987). There remained some, however, who insisted that eugenics contained a "rational core" that continued to merit scientific attention (Paul 1995:125). Several scientist-activists involved in chemical mutagenesis research and in the genetic toxicology movement were among them. James Neel was one student of human genetics who remained adamant that "good" eugenics could be distinguished from "bad" (124). Crow and Lederberg were two others. "How soon and to what extent should man start to intervene in his genetic future?" asked Crow in a 1965 article on genetics and medicine (371). Several of Lederberg's "Science and Man" columns address the relationship between heredity and social progress. In one written shortly after H. J. Muller's death, titled "A Test Tube Daddy," Lederberg (1967) asked his readers, "If genetic betterment can contribute [to societal well-being], how can we refrain from seeking it?" Technical advancements encouraged the view that genetics' main contribution to society lay in improving the gene pool (Sonneborn 1965). By the late 1960s, molecular biology was ripe with promises of future opportunities for "genetic engineering" on a scale never before possible and which no doubt fueled these scientists' barely tempered enthusiasm. "I have one conviction," Crow continued,

> It is high time that the social implications of our genetic knowledge be discussed. Early eugenics was crude, oversimplified, and got confused in various dubious (and in some cases disastrous) political movements. I hope we are ready for a more mature consideration of eugenics and euphenics as complementary possibilities. It may well be that the second century of Mendelism will mark the beginning of a serious and informed consideration of the extent to which man can and should influence his biological future, with full deliberation on both the opportunities and the risks. (1965:372)[25]

Crow's statement supports historian of science Daniel Kevles's (1985) suggestion that eugenics did not so much disappear in the 1960s and 1970s as get reinvented. He and others have argued that the ideological imperative driving the reform eugenics movement was using genetics for human betterment and that this became in the 1960s the leitmotiv of the field of medical genetics. While their medical goals included genetically engineering cures for diseases such as Tay-Sachs or sickle cell anemia, genetic counseling rather than gene manipulation was touted as the most technically and politically feasible eugenics program (Crow 1965; Lederberg 1963a; Muller 1963, 1965).[26]

Despite the widespread sense of vast technological possibility (much of it imagined), the racist backdrop of genetic engineering provided little political opportunity in the era of civil rights for geneticists to push their social educational programs at a national level. In contrast to these geneticists, public enthusiasm for genetic counseling remained muted. "If we really want to change the human population by genetic means, [genetic counseling] is the most likely to succeed. But it is much less clear what society or the individuals comprising it want to do. There is no groundswell of public opinion in favor of doing anything by way of positive eugenics" (Crow 1965:372).[27] Yet at precisely the moment that a social consensus was emerging that reproductive decisions were the responsibility and right of individuals, not society (Paul 1995:129–130), another consensus was emerging that environmental problems increasingly required state intervention. Politicians and federal agency directors responded to the demands of environmental organizations with a series of environmental regulations institutionalizing the state's commitment to environmental protection and regulation (Hayes 1987). These state-sponsored reforms meshed neatly with geneticists' own technocratic interests and recommendations for attacking the problem of chemical genetic hazards, as spelled out clearly in "Chemical Risk" (Crow 1968).

In this context, environmentalism provided an ideological discourse in which geneticists could effectively express their evolutionary concerns. The promotional texts examined in this chapter demonstrate how geneticists' rhetoric interwove their long-held concerns about the social costs of genetic disease with newer concerns about the genetic costs of environmental pollution. Genetic disease and environmentally induced mutations were posed as mirror images of one another: the former provided a concrete picture of long-term decline involving "a general weakening of the species, a borderline state, in which humans become less healthy and the race loses vitality," while the latter provided an explanation of "the consequences of the present load of genetic damage in the human population or the implications, for all of us, if this load of genetic damage were to increase."[28]

The goal of the new eugenics in improving the human genetic condition and the goal of the genetic toxicology movement in preventing further genetic damage were complementary, in both scientific and political terms. The blossoming environmental movement provided the symbolic resources to recast eugenics rhetoric in a language that was palatable to the increasingly race-conscious middle-class public that formed the environmental movement's popular base. And simultaneously, the framing strategy lent a mantle of legitimacy and professional credibility to a new and decidedly lower-status "environmental health" science in a way that would attract geneticists committed to fundamental research. To wit: "Will future generations regard our generation somewhat as we do the pioneers who destroyed our forests and wild life—as geneticists without the wisdom and courage to look to the future?" (Crow 1965:372). Here, the geneticist and the new environmentalist share a common purpose in the protection of nature. The genetic hazards frame ultimately drew its rhetorical energies from the melding of evolutionary concerns well known to geneticists and the environmental movement's preservationist ethic to create a cohesive symbolic package that appealed to biologists and nonbiologists alike.

Conclusion

This chapter has focused on the role of ideas—their packaging and transmission—in the emergence of an interdisciplinary genetic toxicology. Scientist-activists' efforts to redefine chemical mutagens as environmental problems relied heavily on a rhetorical campaign played out in published editorials and essays, public lectures, professional symposia, and congressional hearings. Building the case for genetic toxicology before diverse audiences of potential supporters involved amplifying, extending, and translating a set of collective action frames that described synthetic chemicals in the human environment as *genetic* hazards. Frame amplification highlighted distinctions between causes (chemical and radiation mutagenesis) and between effects (mutagenicity and toxicity) to underscore the empirical significance of environmental mutagens. Frame extension broadened the scope of the genetic hazards frame to encompass an array of disciplinary interests and to identify institutional factors impeding interdisciplinary collaboration. Frame translation linked distinct physiological processes (mutagenesis, teratogenesis, carcinogenesis) to make genetic knowledge and experimental practices indispensable to toxicologists, pharmacologists, and others in the biomedical and public health sciences.

But all the tactical frame alignment in the world would have been useless had the problem of environmental mutagenesis and the solution of genetic toxicology been framed in ways that did not resonate with intended audiences. As a condition of collective action framing, resonance can be particularly difficult to achieve when the empirical credibility of the problem itself is in doubt. This was the situation scientist-activists faced in 1969. While accumulating evidence that environmental mutagens were damaging the genetic material of bacteria, yeast, and fruit flies, there was almost no direct evidence that these same mutagens posed similar threats to humans. To counteract this limitation, scientist-activists grounded their arguments in broader social and political culture, imbuing chemical mutagens with emotionally intense environmental rhetoric. They adopted from *Silent Spring* a narrative structure familiar to environmentalists and an increasingly environment-minded public, casting the problem of environmental mutagenesis as an unintended consequence of scientific progress. Like Rachel Carson's, their story was also one of invisible dangers, urgent needs, and collective resolve leading to societal and scientific redemption. By making connections between the still hidden tragedy of environmental mutagens and several more visible and highly publicized disasters, scientist-activists brought the potential public health threats posed by chemical mutagens closer to non-biologists' experiences with environmental tragedy. And by constructing frames that interwove the geneticists' evolutionary and reform-eugenic concerns with the preservationist spirit of the new environmentalism, their symbolic packaging of genetic toxicology also inspired focused action from mainstream biologists.

6

Organizing a Scientists' Movement

[The EMS] doesn't advertise itself by name as being a place for molecular biology. . . . [I]t is more a green organization, an environmental advocacy group.

—Interview (1997)

An important distinguishing feature in genetic toxicology's historical development, as compared to genetics, biology, or even biochemistry, was the institutional context of its emergence. The basic biological disciplines that arose in the United States in the early twentieth century took their initial form in academic settings—private research universities, land-grant universities, and medical schools (Benson et al. 1991; Kimmelman 1987; Kohler 1982). In contrast, genetic toxicology grew out of the scientific laboratories, advisory boards, and regulatory bodies that made up the federal system of science.

Although momentum was building, chemical mutagenesis circa 1960 remained a highly decentralized genetics subfield. Efforts at interlaboratory coordination were rare, and experimentation was largely ad hoc. Research on what would come to be called "environmental" mutagenesis did not gain an institutional foothold at Oak Ridge until the mid–1960s, with the center of gravity shifting to NIEHS around 1973. These two institutions served as dual anchors for environmental mutagenesis research and test development during genetic toxicology's formative years. Importantly, the field remained anchored in federal science institutions as genetic toxicology expanded in the 1970s, its institutional moorings growing largely *within* the federal health science system—for example, at the EPA, NIOSH, and the FDA's National Center for Toxicological Research. Thus, while university scientists played vitally important roles in genetic toxicology's development from the very beginning, university departments did not.

This had important implications. Disciplines based in university departments propagate and maintain a skilled labor force through their institutional mandate to provide graduate-level education and research training. For sciences based in the federal government, whose missions, generally speaking, are not to train new scientists but to tackle research problems (whether basic or

TABLE 6.1

Environmental Mutagen Societies, by Year Established

Society Name	Year Established
U.S. EMS	1969
European EMS	1970
Japanese Environmental Mutagen Study Group	1971
Environmental Mutagen Research Society (Germany)	1971
Gruppo Italiano Mutageni Ambientali organized as a national section of the EEMS	1971
Section of Environmental Mutagenesis of the Czechoslovak Biological Society formed as a national section of the EEMS	1972
International Association of Environmental Mutagen Societies	1973
Section on Environmental Mutagens formed within the Hungarian Society for Human Genetics	1975
EMS India	1976

applied) that are deemed nationally important, the social reproduction of scientists is a problem whose solution is not nearly so clear-cut.

If, as Robert Kohler (1982:8) and others have argued, university departments function as the organizational building blocks that "embody and perpetuate disciplines," how are disciplines built without them? Put another way, for sciences that emerge out of institutional contexts other than research universities, what social forms accomplish the work of departments? Founding university departments or, for that matter, creating concentrations of research groups within existing departments was not a prominent institutionalization strategy in genetic toxicology; creating new organizations to do the work of departments was. Indeed, it is not overstating the case too much to say that, by and large, organizations organized genetic toxicology.

In the absence of both a technically trained labor force and market-creating legislation (which would be a result, not a cause of scientist collective action), scientist-activists built genetic toxicology from the ground up. They did so by creating mechanisms for recruiting and training scientists, coordinating research, standardizing tools and practices, and public outreach and education. As often as not, these tasks were accomplished collectively, through voluntary associations. Formal organizations played a critical role, as suggested by the formation of no less than nine national or international environmental mutagen societies between 1969 and 1976 (Table 6.1). Less formal committees and ad hoc groups of various sorts also accomplished much of the data gathering, planning, and organizational work required of the movement.

The strategy that evolved, in essence, was a "grassroots" approach to inter-discipline formation. In using this term, I do not mean to imply that the scientists' movement that established genetic toxicology was antiestablishment or that it involved marginalized groups. Rather, interdiscipline building in genetic toxicology was grassroots in the sense that it was built by individuals organized into voluntary associations. Encouraged by the political and structural conditions of their research, scientist-activists created new organizations and modified existing ones. To do so, they often exploited preexisting communication, funding, and research networks—the "mobilizing structures" of the genetic toxicology movement (Tarrow 1998:22–23). They forged new social relations, enrolling chemical and drug companies, U.S. senators, pharmacologists, and research biologists. And perhaps most important of all, they transformed their research tools into research problems by reconfiguring the production logic of mutation genetics. The organization of collective action and the coordination and design of laboratory research fused together. For a short time, at least, doing science and doing politics became one and the same thing. In the United States, that process more or less began with the creation of the EMS.

Alexander Hollaender and the Origins of the EMS

The historical details of the EMS's founding are, for the most part, unremarkable. Much of that history has already been alluded to in previous chapters, and the basic facts are contained in the published record. It is, at root, a story of social connections.

The idea to form a society dedicated to "the study of mutagens in the environment" seems to have originated sometime in 1968 among a small group of "geneticists and toxicologists."[1] This cadre included Alexander Hollaender (Oak Ridge), Samuel Epstein (Harvard Medical School), Marvin Legator (FDA), Matthew Meselson (Harvard), Ernst Freese (National Institute of Neurological Diseases and Stroke), James F. Crow (Wisconsin), Warren Nichols (Institute for Medical Research), and several mutation geneticists working in Hollaender's Biology Division at Oak Ridge, in particular Ernest Chu, Heinrich Malling, and Frederick de Serres.[2] Hollaender was the central node in this social network.

Insider accounts invariably point to Hollaender's oversized personality and his semiauthoritarian management style when assessing his role as a driving force in creating the EMS. There is little doubt that in this and other scientific ventures in which he was involved, Hollaender's charismatic leadership was a factor that must not be overlooked.[3] His social location, professional status, administrative experience, and the subsequent legitimacy he lent to the incipient movement, however, were also critically important.

As noted in earlier chapters, Hollaender was an accomplished and influential science administrator with established links to high-ranking public officials in Washington, the AEC, and the NAS. But his connections were also international. Hollaender was a habitual traveler, as his voluminous travel reports for the Biology Division attest.[4] His excursions to attend scientific conferences, planning meetings, and training symposia frequently took him not only to Western Europe but to Asia and Latin America as well. Indeed, the state of biology in developing countries was one of Hollaender's many pet concerns, and he actively promoted the international exchange of biological information, most notably by organizing the internationally famous Gatlinburg Symposium on Radiation Biology, by bringing dozens of foreign visiting scientists and postdoctoral fellows to study at Oak Ridge, and by organizing symposia and training workshops in developing countries (Setlow 1968:512).[5]

Hollaender's cosmopolitan view of science education paralleled in scale and scope his outlook on the biological sciences more generally. The Big Biology character of his administrative accomplishments are most clearly manifest in the research and training programs he worked to initiate at Oak Ridge.[6] Most notably, the research Hollaender organized at the Biology Division emphasized the study of mutation and nucleic acid chemistry well before molecular biology defined DNA as the primary unit of inheritance (Setlow 1987:1; von Borstel and Steinberg 1996:1052). Other local projects that stand out are the initiation of large-scale mammalian genetics experiments at Oak Ridge conducted in facilities that by 1965 housed some 250,000 laboratory mice (Johnson and Schaffer 1994:114). The establishment of the University of Tennessee—Oak Ridge Graduate School of Biomedical Sciences in 1965 was another local hallmark for which Hollaender could take significant credit.[7]

In addition to his organizing and planning activities at Oak Ridge, throughout the 1950s and 1960s (and beyond) Hollaender was involved in numerous national- and regional-level planning projects in the biological sciences (Setlow 1987:3). For example, he was instrumental in the 1950s in founding the Radiation Research Society (1952) and the Comité International de Photobiologie (1954). In the mid-1960s Hollaender helped organize a NAS steering committee "to plan the future of biology for the next twenty years."[8] And consonant with the establishment of the EMS, Hollaender was working to get biologists involved in studying environmental problems. Projects of this sort, in addition to the EMS, included working to establish regional "environmental science centers" in several major U.S. cities, setting up the Anderson Foundation Lectureship in Environmental Sciences, and working with World Health Organization officials to develop a "World Health Research Center."[9]

In short, by 1969 Hollaender had accrued extensive experience in organizing scientific research and education in a variety of forms. These programs were invariably enabled by and reflect the development of expansive professional networks that included access to funding agency directors inside and outside the federal government. Hollaender not only possessed more than sufficient know-how and social connections necessary for initiating the movement to constitute genetic toxicology, he had access to another highly valuable organizing resource: time.

After stepping down as Biology Division director on January 31, 1966, Hollaender continued to work at Oak Ridge in the position of senior research adviser "with responsibilities for all international activities, symposia, and other outside activities."[10] That is, he was paid to organize scientists.[11] Hollaender's social location could hardly have been better suited for the task at hand. It was Hollaender's social capital, and not simply or only his personal charm or iron managerial hand, that makes him the central figure in this story.

And yet, one may ask, could someone less well positioned than Hollaender have generated the energy and resources necessary to create the EMS and thereby jump-start the genetic toxicology movement? The answer is almost certainly, "yes." If anything, it must be said that, relative to many of his other accomplishments, spearheading the drive to create the EMS was small potatoes. All that was really necessary was the tacit support of enough research scientists to justify the project, a small stipend to cover their limited expenses, and a lawyer to draw up a constitution and bylaws and file for tax-exempt status. A hastily constructed survey on the need for a new society took care of the requirement for collegial approval.[12] An endowment of 110 shares of Atlantic Richfield stock that Hollaender procured from friends at the Anderson Foundation provided the financial backing.[13] And Hollaender's lawyer in Washington, D.C., supplied the legal expertise. Getting the EMS started was simple enough. Growing an interdisciplinary science into something called genetic toxicology, however, presented a different set of organizational challenges.

The EMS as a Boundary Organization

The challenges facing the EMS during its early years were formidable, and the organization consequently wore many hats. Like most formal organizations (Powell and DiMaggio 1991), the EMS prescribed rules for membership, governance, and relations with other organizations and individuals. And like most scientific societies (Whitley 1984), the EMS served as an organ for institutionalizing scientific communication through the publication of a newsletter and the organization of annual meetings. The EMS also went some way beyond these pro forma social functions. As a mechanism for organizing data collection and methods development, for example, the EMS was itself very much involved in constructing the knowledge base and research infrastructure needed to sustain and buttress genetic toxicology. This was accomplished in part by setting up and sponsoring the EMIC and the publication of a ten-volume monograph series published between 1971 and 1986 called *Chemical Mutagens: Principles and Methods for Their Detection*. The EMS also took on a considerable amount of work often relegated to university departments in organizing and sponsoring workshops and short courses in the principles and methods of mutagenicity testing in order to create a trained labor force in genetic toxicology. Amid these various roles and functions swirled an apparent contradiction.

Organizational success depended largely on the level of credibility that the EMS was able to foster among its members, patrons, and consumers of genetic toxicology information in industry, government, and universities and among the general public. Building an interdisciplinary research community, a stable funding base, and a market for genetic toxicology information and practices— all interrelated goals—required the transcendence of partisan interest politics. It became incumbent upon EMS officers and council members to elaborate a vision of genetic toxicology amenable to diverse and often competing interests that were bound to surface at the crossroads of university, government, and industry science; the EMS itself had to remain above politics.

On the other hand, success in the substantive goals set out by the EMS— essentially changing the way mutation research was done, by whom, and for what purposes—relied heavily on overt political rhetoric and action. Scientist-activists sought to challenge the basis of federal chemical regulatory policy by refocusing legislative attention on the genetic impacts of environmental chemicals. As we saw in the preceding chapter, they did so in part by interpreting genetic toxicology in moral and political terms as an issue of scientist, corporate, and government responsibility for the protection of the public health and the preservation of genetic integrity. Encompassing the dual tasks of creating the research infrastructure for the production and dissemination of genetic toxicology information and raising the political saliency of chemical genetic hazards, the EMS functioned simultaneously as a professional scientific society and as an environmental movement organization. How was the EMS able to successfully engage both its scientific and political missions without undermining either?

Recent research on environmental "boundary organizations" offers a framework for making sense of this dilemma (Guston 2001). As defined by David Guston (2000:30), "boundary organizations are institutions that straddle the apparent politics/science boundary and, in doing so, internalize the provisional and ambiguous character of that boundary." Boundary organizations are seen as important institutions in that they create stability across the social worlds of science and politics, in part by facilitating the flow of knowledge or information between them.[14] To the extent that this general model treats science and politics as relatively undifferentiated spheres of social interaction (Miller 2001:483), the EMS provides an opportunity to examine how disciplinary differences within science influence the behavior of boundary organizations.

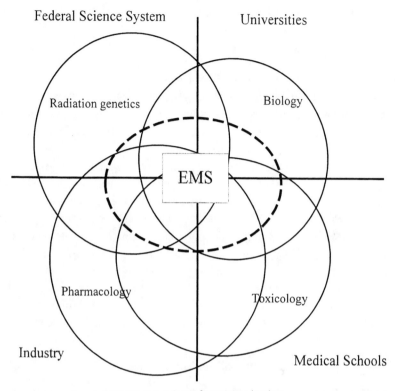

FIGURE 6.1 A Boundary Organization

Emerging as the central organizational actor in the movement to create genetic toxicology, the EMS embodied not one kind of boundary but many (Figure 6.1). Boundaries, whether conceptual, methodological, occupational, disciplinary, or institutional, may be understood in this case as features of the scientific landscape that, from the perspective of EMS scientist-activists, impeded a systematic attack on the problem of environmental mutagens. As one of the central mechanisms created to organize and engage that attack, the EMS did less to clarify and stabilize boundaries than dissolve them. Overcoming taken-for-granted divisions and finding common purpose among disparate knowledge communities were arguably the EMS's most significant achievements. In its capacity as a boundary organization, the EMS regulated the mix of science and politics flowing in and out of genetic toxicology.

Making conceptual sense of these blending processes requires grappling with the complex analytical problem of how different kinds of politics and different ways of legitimating scientific work intersect along different (internalized) boundaries. It also requires that we not assume that all boundaries are created equal but rather that some kinds of boundaries are more important and/or more impervious to crossing than others. In the case of the EMS and genetic toxicology more generally, boundary reordering often involved a two-part strategy in which strict demarcations along certain boundaries were maintained in order to subvert or loosen others. Social boundaries are relational achievements (Gieryn 1999). In order to dissolve some, others have to be reinforced.

To demonstrate how this worked, I present two illustrative examples. The first shows how the EMS specified its area of research interest, setting "environmental mutagenesis" off from other adjacent cognitive and methodological domains. In shoring up the boundary demarcating its scope of research interest, the EMS was able to overcome other barriers keeping geneticists and toxicologists apart. In the second example, I show how the EMS took measures to demarcate a boundary between science and politics, positioning itself on the science side of that line. By establishing a public face as a scientific research society whose task it was to provide impartial information to any and all interested parties, the EMS was able to incorporate a new politics into the production logic of mutation research. In this example, keeping favoritism and environmentalism out of the EMS was a strategy for institutionalizing a new order into mutation research. The resulting reconfiguration of these boundary maneuvers was a thoroughly politicized form of mutation research called genetic toxicology.

Controlling Mutagenesis

According to Kelly Moore, boundary organizations "provide both an object of social action and stable but flexible sets of rules for how to go about engaging with that object" (1996:1598). The EMS was quick to demarcate its scope of research interest by defining "environmental mutagenesis" as its object of action. It proved to be a useful strategy for negotiating several boundaries.

By 1969, "mutagenesis" described a relatively well defined set of genetic processes at the empirical center of the newly emerging field. Mutagenesis also defined an important part of the genetics material culture, resembling what Joan Fujimura (1996:5) calls "theory-methods packages." These consisted of a set of interconnected biochemical theories about how and why chemicals and radiation induce genetic mutations and mutagenicity bioassays and experimental protocols for "realizing, materializing, testing, exploring, and adjusting the theory." These packages structured scientist-activists' relationships to the material culture of mutation research, to one another, and, on occasion, to policy makers. Thus the term "mutagenesis" implicated an entire menu of loosely connected practices, concepts, materials, technologies, and people—geneticists and toxicologists and sometimes legislators—for inducing and interpreting mutational change in laboratory organisms and for extending that knowledge into other disciplines.

To better appreciate the important role that the EMS played in renegotiating disciplinary boundaries, it helps to recall that at the time there were no inherent technical or theoretical reasons why EMS Council members chose to restrict its focus on chemical effects research to the genetic level. Nothing prevented the EMS from adopting a different name to reflect a broader approach. Indeed, some members of the EMS Council favored it. Suggestions to widen the scope of the society to include "environmental effects other than the purely genetic," or "chronic biological effects," or, more specifically, teratogenicity and carcinogenicity came up in several discussions during EMS executive, business, and council meetings in 1970. During these conversations, however, council members voiced concern that "such expansion might dilute the genetic thrust of EMS," and the issue was repeatedly tabled.[15] As this quote suggests, restricting the EMS's field of interest to mutagenicity involved more than simply taking advantage of the then-available concentration of expertise in mutation research. At stake was the integrity of the mutagenesis theory-methods package that provided the methodological rules for engaging environmental mutagens. As Hollaender urged in one essay written in 1970, "The methodology does exist to solve the problems of environmental pollution. All that is needed is coordination among scientists and institutions and sustained financial support."[16]

The decision to adopt the qualifying term "environmental" as a means of identifying both the new society and the field of research it represented implicated at least three other boundaries. First, the term effectively politicized mutation research. In interviews, scientists admitted that the decision to incorporate the word "environmental" into the name of the society was an ad hoc one based mainly on the recognition that the environment had become a hot-button issue. "It was a fashion. It was one of those trendy words," one scientist told me.[17] But it was also a political strategy. In an era of rising ecological awareness, enrolling the environment was a way to attract attention and, hopefully, support. Second, the term distinguished work promoted by the EMS from an older school of mutation research that was concerned strictly with mutagenesis as a problem of basic biology. As opposed to mutation research, environmental mutagenesis emphasized those substances, mostly chemical in nature, that were present in the human environment and thus posed a potential risk to the population or, in the case of occupational exposures, to specific subpopulations. Third, the array of potentially mutagenic substances that fell under the purview of the EMS expanded considerably, from those few chemical classes shown earlier to induce interesting mutations in laboratory organisms to the tens of thousands of substances released into and circulating through the environment. In that sense, the problem domain described by the EMS became, by definition, an interdisciplinary one. Taken to its logical conclusion, research on the genetic risk of environmental chemicals could be interpreted as requiring the expertise of analytical chemists and perhaps even ecologists to understand the formation, bio-concentration, persistence, photochemical and metabolic transformation, and long-range transportation of mutagens through air, water, and soil (Fishbein 1973). Conceptualized thus, environmental mutagenesis contained a broad set of problems that genetics alone could not capably solve.

Scientist-activists' decision to enroll the environment and simultaneously retain a strict focus on genetic effects reveals some of the disciplinary and political stakes involved: the EMS needed to attract an interdisciplinary membership and audience, but not at the expense of relinquishing control over the domain of knowledge, skills, and techniques that motivated the research questions and made the science "doable." Germ-cell mutagenesis was the EMS's trump card. It was the one biological process in which no other field of environmental health science could claim any demonstrable expertise. In that sense, mutagenesis was bound up tightly with geneticists' authority. But controlling mutagenesis was not simply a tactic for maintaining professional control in an expanding interdisciplinary field. Had it been, scientist-activists would have been far more reluctant to share their knowledge with others than they actually were. But as we will see, rather than hoard their skills, these scientists led the charge to train a new labor force of genetic toxicologists.

The tools and techniques of mutation research were political weapons in a much broader sense than disciplinary ethnocentrism allows. The mutagenesis theory-methods package offered a means of establishing a measure of certainty in an area of science where uncertainty reigned. The political value of mutagenicity bioassays was manifest in their potential to demonstrate the genetic risks associated with chronic chemical exposure to current and future generations. In that, "environmental mutagenesis" represented an arsenal that no other field of biology could effectively wield in the ongoing battle to tighten regulatory control of chemical production and use.

Lowering Institutional Barriers

Where the stabilization of a cognitive and technical domain had required boundary maintenance, attracting scientists and patrons from a variety of disciplines and across research sectors required partially dissolving other boundaries. This would not happen automatically; the stabilization of interdisciplinary relations was something that had to be achieved through concerted collective effort.

In an attempt to mobilize as broad a pool of potential activists as possible, exceptionally low membership dues were instituted ($10 annually), and membership requirements were relaxed at the first opportunity. An amendment to the original bylaws, instituted in the spring of 1969, extended membership to "scientists *and other qualified persons* who share the stated purpose of the

Society and who have relevant knowledge and experience."[18] The change was small but significant. It meant that a graduate degree in genetics was not a prerequisite for EMS membership. This permitted formal participation by students, scientists, and technicians without specific training in genetics and administrative personnel from industry and government. Lowering the bar to membership proved an effective organizational strategy. Membership in the EMS jumped from 87 in June 1969 to 452 a year later.

Having made joining the society possible for just about anybody, the EMS launched an extensive publicity campaign to give researchers across the biomedical, biological, and agricultural sciences access to information that might encourage them to join. At the first EMS meeting, a "public contact committee" was created, and the newly elected newsletter editor was told to "make sure the Newsletter circulates to toxicologists, people interested in carcinogenicity, and public health people."[19] Announcements were published in journals ranging from the general (e.g., *Science*) to the specialized (e.g., *Chromosoma*), from the basic (e.g., *Journal of Heredity*) to the applied (e.g., *Journal of Industrial Hygiene*), and included major biomedical journals such as *Cancer Research*, *JAMA*, and *Journal of Nutrition*, among many others.[20] EMS members were encouraged to submit papers dealing with environmental mutagenesis to journals read by toxicologists and pharmacologists rather than the genetics journals that traditionally published these kinds of studies.[21] Several EMS members promoted the new field in speaking engagements in the United States and Europe. Many of these were lectures or talks given before other scientists, but some, like Heinrich Malling's lecture, "Chemical Mutagens in Our Environment," were given to nonprofessionals, in this case an undergraduate botany class at the University of Tennessee.[22]

An important aspect of the campaign to attract interdisciplinary audiences involved establishing relations with other organizations. To that end, EMS committees and individuals took an "active part in the deliberations of the Weed Society, Society for the Study of Aging, Pesticide Board, Drug Research Board, and [advising] Congressional Committees."[23] They also worked to establish more formal liaisons with the Radiation Research Society, the Society of Toxicology (SOT), the Genetics Society of America, the International Teratology Society, and others. Often this involved arranging for overlapping meetings with jointly sponsored symposia. "Possible contributions of genetics to toxicology," for example, was the topic of one such panel convened jointly by the EMS and the SOT in 1971.[24] The program at the fifth EMS annual meeting included an EMS-sponsored symposium titled "Evaluation of Mutagenicity Data and Its Toxicologic Significance," another sponsored by SOT titled "Nitrosamines and Nitrosamides: Environmental Occurrence and Toxicological Significance," as well as a regular session titled "Hycanthone and Its Derivatives: An Integrated Approach to Environmental Mutagenesis."[25] Panel discussions on topics related to genetic toxicology also were organized by EMS members at meetings of the Radiation Research Society (in 1970).[26] Attempts at organizing a similar panel at the International Congress of Human Genetics (in 1976) were less successful but serve to buttress the broader point that an intensive boundary-crossing effort to establish intersocietal relationships was a major function in EMS during its early years (Armendares and Lisker 1977).[27]

Another set of important organizational actors were the chemical and drug companies. Initially, they were enrolled by EMS officers as financial sponsors for EMS conferences. Hollaender and EMS treasurer Marvin Legator wrote letters to pharmaceutical companies requesting contributions to finance the first two EMS conferences with considerable success; chemical companies contributed roughly $6,000 to the second (Table 6.2).[28] It was not long before industry scientists themselves were courted in earnest. Hollaender's not uncontroversial opinion on the issue was that if industry were actively involved in EMS from the beginning, there would be less resistance from industry when the push for stiffer regulations came around.[29] From the firms' point of view, in the face of stepped-up regulation, short-term mutagenicity bioassays were far less expensive than carcinogenicity tests. If a chemical substance was harmful to the public health, it was far better to know about it before the compound went into full development than after.

In a 1982 interview, Hollaender recalled that some academic scientists had complained of the industry presence in EMS. This would have been expected given the long-standing economic and ideological divisions separating academic- and industry-based science. My examination of archival evidence, however, revealed little protracted conflict during this early period over the organization's inclusion of industry scientists. It must have been fairly clear to those involved that the substantive content of genetic toxicology represented the merging of basic and applied approaches to the problem domain defined by environmental mutagenesis. New mutagenicity bioassays could not be developed without understanding how mutations are created. By the same token, new methods would and did lead to insights into more fundamental questions about the nature of mutation. As discussed earlier, this relationship between new methods and new knowledge had always existed in genetics. The difference was that in 1970, a movement to routinize these mutually reinforcing production processes and a new set of actors—chemical and drug companies—were playing increasingly important roles in institutionalizing those changes. As a boundary organization, the EMS itself was rendered permeable to industry influence. The makeup of its executive council reflected the increasing convergence of university, government, and industry research sectors in genetic toxicology. In 1974 EMS members elected two scientists employed by for-profit firms to executive office, and by 1976 the editorial boards of journals such as *Mutation Research, Environmental Mutagenesis and Related Subjects, Genetic Toxi-*

TABLE 6.2

Corporate Sponsors for Second Annual EMS Conference, 1971

Firm Name	Source	Amount
Hoffman La-Roche	Burns to Legator 12/15/70	$500
Warner-Lambert Research Institute	Briziarelli to Legator 1/11/71	200
Schering Corp.	Tabachnick to Legator 1/5/71	500
Smith Kline & French Laboratories	Buckley to Legator 1/12/71	300
Arthur D. Little Co.	Hollaender to Kensler 2/24/71*	250
Ciba Pharmaceutical Co.	Diener to Hollaender 2/16/71	200
Merck Institute for Therapeutic Research	Peck to Legator 1/7/71	250
Diamond Shamrock Chemical Co.	Eisler to Legator 1/19/71	500
Pfizer Pharmaceuticals	Ray to Legator 2/5/71	500
Procter & Gamble Co.	Kennedy to Legator 12/23/70	500
Dow Chemical Co.	Hollaender to Kilian 2/24/71*	500
Abbott Labs.	Philip to Legator 3/5/71	250
Bristol Labs.	Gardier to Legator 12/1/70	0
Mason Research Institute	Mason to Hollaender 12/3/70	0
Geigy Pharmaceuticals	Barclay to Hollaender 12/18/70	500
Ciba Pharmaceutical Co.	Diener to Hollaender 2/16/71	200
Cutter Laboratories	Guzman to Hollaender 12/3/70	0
General Foods Corp.	Kirschman to Hollaender 12/9/70	0
Squibb & Sons	Smith to Hollaender 12/7/70	500
USDA	Klassen to Hollaender 12/9/70	0
Sandoz Pharmaceuticals	Timms to Hollaender 1/27/71	0
Cutter Laboratories	Guzman to Hollaender 12/3/70	0

Source: This table was constructed from correspondence contained in the folder "Solicitations for support of first EMS meetings," EMSA (the folder is mislabeled).

*Letters from EMS representative acknowledging receipt of gift.

cology Testing, and *Reviews in Genetic Toxicology* reflected the increasing participation by industry scientists.

The Public Face of the EMS

Scientists troubled by potential genetic hazards littering the human environment did not have to join the EMS. Neither the nature of the problem of environmental mutagenesis, as it was then understood by genetic toxicology scientist-activists, nor the intensive publicity campaign they pursued is sufficient to explain the rapid, sudden, and heterogeneous character of mobilization. Organizational credibility was also paramount to movement success.

The fact that the founding members of the EMS were for the most part established scientists and science administrators no doubt helped gain the early attention of elites in Congress and the NAS. But the legitimacy of the EMS—and the movement that emanated from it—depended ultimately on a generalized perception of the organization's impartiality, a spirit of scientific neutrality that not only accompanied official statements but also was embodied in the routinized activities of the organization and its various interdiscipline-building projects. To attract members and financial backing, the EMS had to present itself as a scientific research society committed first and foremost to the production, rationalization, and dissemination of objective knowledge. To do so, the EMS strove to keep environmental politics out of genetic toxicology data and information.[30]

The public face of the EMS is perhaps best described in a "statement of activities" contained in an Internal Revenue Service report filed on behalf of the EMS in 1969. "Like most organizations of scholars," the report read, "the EMS will, through scientific congresses, symposia, a journal and a newsletter, provide the traditional forums through which scientists of similar professional interest have for generations communicated with one another and with the public. Experimental data and new theories are shared and subjected to the inspection and critical review of informed colleagues."[31] The report, which sought to establish tax-exempt status for the organization, presented an idealized description of the EMS as a scientific research society and therefore should not be taken as evidence of the full range of activities and functions the EMS served in practice. More important is that tax-exempt status placed definite constraints on the kinds of political activities the EMS could legally pursue.[32] These legal constraints enhanced the organization's credibility as one whose main, and perhaps only, formal interest was in the "inspection and critical review" of scientific knowledge.

On that basis, formal relations with environmental groups, for example, were roundly discouraged. In reference to a letter that Joshua Lederberg received from the Natural Resources Defense Council, purportedly requesting information on environmental mutagenesis and which Lederberg brought to the attention of the EMS Council, EMS president Hollaender "proposed to make it clear . . . that EMS would be willing to function only as a resource facility, and not in the development of any action program."[33] Five years later, Hollaender complained again of frequent requests for information on "chemical toxicology," this time from the group Resources for the Future.[34] I found no evidence, however, that in the interim the EMS entered into any kind of relationship—formal or otherwise—with environmental organizations.[35]

Organizations that may be assumed to have harbored political and economic interests biased in the opposite direction received similar rebuffs. A proposal that the Association of Analytical Chemists be invited to review validity and reproducibility studies of mutagenicity tests was struck down on the grounds that the association "has no special expertise in this matter."[36] The same attitude guided relationships with firms having a direct economic interest in the production of genetic toxicology data. A report from the EMS Committee on Methods advanced the position that "the EMS should avoid putting itself into a position of certifying or providing an endorsement to any laboratory or test method. It should serve only as an assembly of scientists willing to provide individual expertise, upon request, to anyone requesting it."[37] The committee advocated this position as a means of avoiding potential legal difficulties or conflict of interest charges. Such outcomes would threaten the EMS's appearance of organizational neutrality and undercut efforts by the EMS leadership to institutionalize ideological purity. Those efforts are perhaps best illustrated by considering the science/politics boundary as it came to be embodied in EMIC.

Spreading the (Objective) Word

EMIC was a direct outcome of the creation of the EMS.[38] Housed at Oak Ridge and initially directed by Heinrich Malling, EMIC served as an information clearinghouse for mutagenicity data. It employed a small technical staff charged with collecting published literature on chemical mutagenesis, condensing the data presented in those articles into uniform tabular abstracts, and building a computer database from that information that could be accessed via one of a number of standardized index codes as needed for specialized literature searches (Malling 1971; Malling and Wassom 1969).

EMIC's other primary task was disseminating that concentrated information. The main mechanism for this was an annual literature survey that EMIC produced and distributed, mostly to members of the various EMS societies around the world.[39] EMIC staff also published occasional "awareness lists"— short bibliographies of important subclasses of chemical compounds—in the *EMS Newsletter*. In addition to these general means of information dissemination, EMIC staff attended to the specific requests for data of "anyone who requested it."[40] "Anyone," in this case, seems to have been interpreted quite literally. Indeed, keeping not only scientists but also "the general public informed about highly technical data" was a central concern and explicit function of EMIC.[41] In 1972, for example, EMIC staff reported receiving 222 individual requests for information. "The greatest proportion of these requests were from persons engaged in research, but some came from a variety of sources," the report noted. These included "city municipalities, high school students, free lance writers" and the occasional legislator.[42] Malling's official correspondence record for 1970 confirms both that EMIC staff spent a considerable amount of their time handling information requests from outside and that a nontrivial proportion originated with concerned citizens from various walks of life.[43]

The available evidence does not reveal whether EMIC staff actually responded to all of these information requests. In light of the continuous stream of complaints regarding inadequate staff and funds, probably not. One might reasonably assume that under conditions of resource scarcity, requests from high school students or citizens' groups might not receive the same level of attention as those coming from scientists active in the mutation research field. What is clear is that, in principle, EMIC—and by direct extension the EMS—was committed to serving the public interest as an impartial messenger of genetic toxicology information. That impartiality extended into the economic sphere as well: although the issue was raised several times at EMS meetings as a means of offsetting tight budgets, neither industry scientists nor foreign scientists (nor any other groups for that matter) were charged fees for EMIC services.[44]

Institutionalizing Impartiality

In short, the EMS was keenly interested in disinterestedness.[45] The society's leadership pursued policies of conduct, scientific review, and public service that depended on and reinforced a strict division between the EMS as a scientific research organization and politics of various stripes—from environmental protest to the endorsement of particular testing protocols. The organization's public boundary work (Gieryn 1999), embodied most explicitly in the social service functions of EMIC, can be understood as an explicit attempt to establish a very distinct, unyielding, and *publicly visible* boundary around the EMS, EMIC, and, by extension, genetic toxicology.

It would be wrong, however, to interpret these policing efforts as only or merely ideological in nature. They were also born of organizational necessity. At the time, genetic toxicology could boast few if any stable sources of funding. As Heinrich Malling wrote to Frits Sobels in the fall of 1970, "The money situation in the U.S. is very tight. There is essentially no money for screening for the mutagenicity of harmful pollutants. Besides the standard mutagens such as EMS, MMS, etc., the only new compound with which research is in progress is cyclophosphamide."[46] Much of the funding at the time came in the form of budget line items during a period of general decline in the funding rate for basic research. Money for EMIC and for other EMS projects was not at first easily obtained or readily recommitted. Numerous federal agencies, various chemical and drug companies, private foundations, the National Laboratories, and four or five of the National Institutes of Health contributed small sums to sponsor EMS conferences and workshops and to support the work conducted at EMIC, usually on a year-by-year basis.[47] Given the heterogeneity of these patrons, and the resulting instability of an economic foundation underlying research and development in genetic toxicology, the EMS's boundary-making activities were an organizational survival strategy and not merely an ideological reaction to "politics," environmental or otherwise. What's more, it was a strategy that worked.

As in studies of other boundary organizations (Keating 2001), the institutionalization of organizational impartiality served a number of specific practical purposes. It helped secure EMS's tax-exempt status, which reduced the young organization's economic burden even as it reinforced the science/politics boundary through restrictions on political lobbying and partisan endorsements. The rhetorical construction of "good" science in the interest of environmental health also facilitated the enrollment of scientific and political elites. Conversely, the same spirit of neutrality gave drug and chemical companies little room to charge the EMS with environmentalist bias and therefore avoid taking some responsibility in funding and participating in its development.

It is difficult to imagine the same kind of support coming from so many different quarters if the EMS had not made the focused efforts it did to draw this boundary, but the experience of the Japanese Environmental Mutagen Society (JEMS) may be an indication of the organizational costs of not maintaining some ideological distance from environmental politics. A representative of that society reported to EMS Council members in 1972 that "a major problem at the development of [JEMS] had been political implications on environmental problems, as 'left-wing' parties were using these issues to attack the Japanese government."[48] At the same time that JEMS was being hampered by its association with the environmental movement in Japan, in the United States the EMS was quickly gaining firmer financial and organizational footing. After two years of very precarious budgeting arrangements, in 1972 both the FDA and NIEHS committed large sums to support EMIC. With this support, these federal agencies gave the EMS and EMIC a stamp of legitimacy and provided a public endorsement of the importance of genetic toxicology research.[49] By 1977, NIEHS was funding EMIC to the tune of $190,000 per year (NIEHS 1977:331).

Maintaining a strict distance from groups with clear-cut political and/or economic interests represented what sociologists of science have come to recognize as a major component of boundary work: negotiations that seek to balance the competing claims that science is at once socially relevant and unblemished by social bias (Gieryn 1999; Kleinman and Solovey 1995; Moore 1996). Creation and maintenance of that boundary, in turn, enhanced the legitimacy of the EMS's overall project—genetic toxicology—as well as the organization's own credibility and autonomy as a boundary organization. The measures the organization took to routinize its relations with stakeholder groups helped it gain access to resources (elite support, money, and institutional space in the emerging environmental state) and to secure a measure of control over the direction of genetic toxicology research, testing practices, and policy recommendations.[50] As the one scientific body that maintained a relative monopoly on the aggregation and distribution of environmental mutagenesis knowledge, the EMS would come to possess considerable influence. Keeping environmental politics out of genetic toxicology was ultimately a strategy for gaining and holding on to that authority. It was also, paradoxically, a strategy for getting the EMS's own political work accomplished inside genetic toxicology.

Fusing Research and Activism

The genetic toxicology movement created a new market for scientific knowledge. As a boundary organization that regulated the flow of people, knowledge, and money into and through this emerging field, the EMS was instrumental in setting these market forces in motion. To understand how, it is useful to recall the distinction I drew earlier between the contentious collective action that is the basis of many social movements and those conventional modes of collective action employed by scientists' social movements. The case of genetic toxicology does well to illustrate the latter. Working collectively through the EMS, scientist-activists succeeded in routinizing a set of new and modified research practices that described the implementation of a new order of environmental inquiry. In effect, the EMS became its own environmental movement organization. At least for a time, science and scientist-activism merged. Doing experiments, collecting and disseminating data, and building the political and moral case against the indiscriminate use of mutagenic chemicals were tightly intertwined and mutually constitutive strategies. The organization's material and ideological commitments—to the production of accurate knowledge, to the integrity of the new discipline, and to the reduction of environmental genetic hazards—converged and were manifest in three modes of conventional collective behavior: committee work, curriculum development, and experiment design.

Activism by Committee

The role of the EMS in organizing genetic toxicology has by now been well established. But the EMS represents only one form of organization that mattered. Committees and subcommittees, although often less formal and less enduring over time, undertook much of the behind-the-scenes work of building the new discipline. The EMS, in fact, was carpeted with committees. Some were those common to most professional societies, for example, a planning committee, a membership committee, a committee to handle publication of the newsletter, and another to deal with "public contact." Other committees reflected some goals specific to EMS. Minutes from the first formal meeting of the EMS leadership indicate the formation of a committee for establishing a chemical registry (what would become EMIC) and another for looking into the publication of a monograph on mutagenicity testing methods. These committees embodied strategies for either meeting the basic requirements of any scientific society or the achievement of specific organizational goals. In that, they were unremarkable.[51]

Other EMS committees, however, took on tasks that bore implications far broader than its own organizational survival or identity and had fairly explicit and direct implications for the direction of research in genetic toxicology and chemical regulatory policy. Some consisted of teams organized to review literatures specific to a particular mutagen or mutagen class. Examples from just the first few years include committees to study the mutagenicity of caffeine, cyclamate, mercury, hycanthone, and nitrosamines.[52] Some were organized to establish recommendations for standard protocols for mutagenicity testing. In 1970 a "methods committee" was appointed "to critically assess recommended methodologies and also to recommend and evaluate future research and method development." The "cytogenetics committee" was another, more specific methods committee established the following year to essentially conduct the same set of tasks with respect to in vitro human cell tests that could be used to screen for chromosome aberrations in humans.[53] The latter was in specific response to questions about the correlation between human genetic risk and positive mutagenicity in bioassays relying on submammalian systems. Still other committees were more explicitly political. One committee, created in 1972, was charged with "extending [the Delaney Clause] to mutagens and teratogens," and another, created in 1975, looked into issues of chemical protection for workers.[54]

These committees were formed within the EMS to accomplish very concrete tasks and upon their completion were generally disbanded. In that respect, they resembled the kinds of "action committees" common to social movement organizations more so than the specialty sections or interest groups commonly found in larger professional societies. Advisory committees in particular served important boundary-ordering and boundary-crossing functions. By reinforcing divisions between experts and nonexperts even as they simultaneously redirected mutation research toward issues and questions relevant to environmental health, advisory committees sponsored by the EMS helped to directly shape science and regulatory policy choices.

Such was the "Committee 17" report. As its name suggests, this was a seventeen-member scientific review body convened by the EMS Executive Council to review research in mutagen detection and in population monitoring, assess the risk implications of the data derived from these testing and monitoring systems, and recommend directions for future research and for chemical regulatory policy. The committee's final report was published in *Science* in 1975 under the title "Environmental Mutagenic Hazards" (Drake et al. 1975). It was essentially the EMS's position paper on research needs and regulatory responsibility for managing chemical genetic hazards. The report argued forcefully that "it is crucial to identify potential mutagens *before* they can induce genetic damage in the population at large" (504). Human protection from environmental mutagenic insult, however, required both the development and validation of new methods for detecting environmental mutagens and the creation of a trained labor force to design, conduct, and interpret those tests.[55]

Pedagogical Activism

Upon its emergence, the scientists' movement to establish genetic toxicology faced an immediate and acute shortage of people trained in the principles of chemical mutagenesis and genetic toxicology testing methods. "Although methods are presently available for evaluating mutagenic agents, and the demand for such testing is increasing, there is a critical shortage of trained personnel," Marvin Legator noted with some urgency. "Industry, government agencies, and universities are seeking trained biologists who are familiar with some or all of the proposed methods for mutagenicity testing. There is no center where individuals can receive formal instruction in this area."[56]

University graduate programs at the time seem to have provided few of the immediate conditions necessary to quickly raise a new and proficient labor force. While the study of mutation and mutagenesis was an established part of most genetics curricula (undergraduate as well as graduate), only those students working with scientists directly involved in environmental mutagenesis research would have been introduced to the latest developments in mutagen testing (Straney and Mertens 1969).[57] Moreover, Ph.D. candidates in genetics almost certainly would not have had training in toxicology or pharmacology, unless specializing in those areas. Graduate students in toxicology, itself a newly emerging field still very much tied institutionally to medical school pharmacology departments (Hays 1986), also would not have been introduced to genetics except perhaps to fulfill general requirements in biological sciences. Specialized studies in genetics were not a common feature of the toxicology Ph.D. In general, there were no ready-made niches for genetic toxicology in university science.

Training in genetic toxicology by necessity developed eclectically, through a variety of mechanisms, most of which directly involved people associated with Oak Ridge, NIEHS, the FDA, and/or the EMS. One critically important training mechanism was the postdoctoral fellowship. Over time, dense research and teaching networks linked some university graduate programs in biology or genetics to these federal laboratories. If those postdoctoral trainees continued in environmental mutagenesis, chances were good that they remained in the genetic toxicology institutions emerging within the federal science system. An annual report issued in 1975 from the Environmental Mutagenesis Branch at NIEHS noted that "the great shortage of scientists to do research in the area of environmental mutagenesis both in the United States and abroad has provided incentive for staff scientists to develop a training program at both the predoctoral and postdoctoral levels" (NIEHS 1975a:155).[58]

In the interim, the EMS and, after 1973, the International Association of Environmental Mutagen Societies (IAEMS) played important roles in providing crash courses in mutagenicity testing methods, bioassay design, and genetic principles underlying the tests. A three-day "Workshop on Mutagenicity" convened at Brown University in the summer of 1971 was the first of several mini training sessions organized to introduce geneticists, toxicologists, and medical researchers to available testing and evaluation methods. The workshop's sponsors included the EMS, FDA, NIEHS, Drug Research Board, and Pharmaceutical Manufacturers Association Foundation. Four main panels were organized around different testing methods: host-mediated tests using bacteria and yeast, host-mediated tests using mammalian cells, the dominant lethal test (mouse), and tests for chromosome breakage. NIEHS director David Rall delivered the plenary address on the topic of the "role of environmental problems in clinical medicine"; Charlotte Auerbach and Alexander Hollaender took part in the summary panel. A similar workshop, also sponsored by the EMS, was convened the following year in Zurich, Switzerland.[59]

In 1972 the EMS Executive Council passed a resolution to establish "a comprehensive workshop on procedures for detection of chemically induced mutations." The proposal for that workshop, drafted by EMS treasurer Marvin Legator, called for monies to support a month-long workshop that would include a "series of lectures and intensive laboratory instruction to develop biologists with a working knowledge of principles in this field." The main tests and techniques taught at the course would include cytogenetic tests for chromosome breaks, dominant lethal studies in rats and mice, host-mediated assays, and bioassays using microorganisms. Legator's tentative course outline involved morning lectures, afternoon laboratories, and evening working sessions on topics such as cytogenetic slide reading, interpretation of statistical data, and the integration of mutagenicity tests into standard toxicological screening programs. The proposed nine-member organizing committee included seven EMS members.[60] Although funds for this workshop did not materialize, smaller week-long workshops did continue to be offered fairly regularly in Italy, Canada, Great Britain, India, and the United States.[61] In conjunction with a workshop on "detection of environmental mutagens and carcinogens" held in 1976 at the University of Texas Medical Branch in Galveston, five participants from countries in Asia and Latin America represented a first step in promoting genetic toxicology training in Argentina, Brazil, Egypt, Japan, and Mexico.[62]

Research Activism

The genetic toxicology movement redefined chemical mutagens as genetic hazards and in so doing set in motion a series of processes that in a relatively short period reconfigured how mutation research was conducted, by whom, and for what purposes. In accomplishing this, the movement was forced to confront two critical issues: how to identify environmental chemical mutagens and how to estimate the genetic risk they posed to humans. These issues were interrelated and mutually reinforcing. One of the keys to understanding how genetic toxicology became institutionalized hinges on appreciating how scientist-activists responded to these intertwined challenges to established mutation research practices. The result was a transformation in the genetics political economy.

Prior to the rise of the genetic toxicology movement, geneticists and biochemists familiar with chemical mutagens saw them as tools designed for use in a highly specialized field of genetics research. Within the space of a few short years, that general understanding ruptured as chemical mutagens came increasingly to represent the seeds of an environmental health problem of unknown proportions. The new meanings that the movement attributed to chemical mutagens effectively inverted the logics of production that had governed mutation research practices for nearly three decades.

Where before scientists had focused their attention on a handful of highly potent chemical mutagens, because either their modes of action or their mutational effects offered insight into genetic-level phenomena, scientist-activists now focused increasingly on the broad spectrum of potentially mutagenic chemicals and their attendant environmental risks. Where experimental design had once emphasized a few theoretically interesting chemical mutagens, production goals in genetic toxicology, shaped by the movement's commitment to prevent an increased frequency of human genetic disease, came to emphasize the rapid identification of as many chemical mutagens as possible. This shift in logic favored mutagenicity bioassays that were quick, inexpensive, unambiguous, and sensitive enough to screen even very mild mutagens since chronic and possibly synergistic effects were suspected but largely unverified. Historically, geneticists interested in questions about gene structure and function had been attracted to research organisms whose physiology and behavior complemented those research interests. As noted earlier, *Drosophila* was well suited to early studies of gene mapping in part because it possessed exceptionally large chromosomes that were visible under standard laboratory microscopes (Allen 1975). Bacteria and the bacteriophage viruses also became popular laboratory animals due in part to their ability to withstand moderate levels of mutagenic insult and to reproduce rapidly by the millions (Drake 1970). These and other bacteria, yeast, fungi, plant, and insect bioassays were readily available, and scientist-activists used them in efforts to identify environmental mutagens efficiently and rapidly.

As many soon realized, however, the right tools for studying the mechanisms of inheritance were not necessarily the best tools to produce realistic estimates of the genetic risk from human exposure to chemical mutagens (Clarke and Fujimura 1992). The problem of extrapolation threw a wrench into the production logic that had earlier guided the development and standardization of bioassays in mutation research. As geneticists had suspected since the 1950s, mutagen specificity is influenced by metabolic processes of living organisms (Lederberg 1997). "Some pollutants in the environment are neither mutagenic nor carcinogenic by themselves but can be converted by mammalian metabolism to highly reactive and genetically active metabolites. Microorganisms have only a fraction of the toxification-detoxification mechanisms that a mammal has" (Malling 1977:263). Scientist-activists' interest in generating experimental results that could reasonably be generalized to humans favored mutagenicity bioassays that took mammalian metabolic processes into account. Mutagenicity bioassays that used mammals were very expensive and very time consuming, and compared to the number of existing submammalian tests, there were far fewer from which to choose.

These two critical needs—to quickly and accurately identify environmental mutagens and to estimate the genetic risk associated with them—defined the basic parameters for a new economy of genetic toxicology practices that would emerge in the early 1970s. As scientist-activists sought to design test systems that optimized requirements for speed and sensitivity, on the one hand, and generalizability to humans, on the other, a hotly competitive test development economy emerged. As it did, considerable effort was put to adjudicate among these competing systems (Table 6.3).[63]

Two general strategies defined the area of most intense competition. One sought to create in vitro bacterial systems that incorporated humanlike metabolism. These were the microorganism bioassays with metabolic activation, of which the so-called Ames test is most renowned (Ames 1971). The other sought to create in vivo mammalian systems that incorporated bacterial indicators. These were the "host-mediated" bioassays in which bacterial cultures were surgically implanted in live animals, who were then treated with chemicals, sacrificed, and the bacterial cultures examined for mutagenicity (Legator and Malling 1971). Besides these two general strategies that mark the center of a swirling storm of activity, scientist-activists designed and promoted numerous other systems, techniques, and methods for testing compounds for various mutagenic endpoints. The 111 articles that were published between 1971 and 1986 in the ten-volume series *Chemical Mutagens: Principles and Methods for Their Detection* (Hollaender 1971a,b, 1973, 1976; Hollaender and de Serres 1978; de Serres and Hollaender 1980, 1982; de Serres 1983, 1984, 1986) describe about half of the roughly 200 test systems estimated to have been developed during that period.[64] These test systems, some of which are still in use although many have

TABLE 6.3

Operational Characteristics of Mutagen Screening Systems

Test System	Time to Run Test	Operating Costs	Initial Investment Costs	Relative Ease of Detection	
				Gene Mutations	Chromosome Aberrations
Microorganisms w/ metabolic activation:					
Salmonella typhimurium	2–3 days	very low	low	excellent	
Escherichia coli	2–3 days	very low	low	excellent	unknown
Yeasts	3–5 days	very low	low	good	good
Neurospora crassa	1–3 weeks	moderate	moderate	very good	
Cultured mammalian cells w/ metabolic activation	2–5 weeks	moderate to high	moderate	excellent to fair	unknown
Host-mediated assay w/					
Microorganisms	2–7 days	low to moderate	low to moderate	good	
Mammalian cells	2–5 weeks	moderate to high	moderate	unknown	good
Body fluid analysis	variable	variable	low to moderate	variable	
Plants:					
Vicia faba	3–8 days	low	low		relevance unclear
Tradescantia paludosa	2–5 weeks	low to moderate	moderate	potentially excellent	

Test System	Time to Run Test	Operating Costs	Initial Investment Costs	Relative Ease of Detection	
				Gene Mutations	Chromosome Aberrations
Insects: *Drosophila melanogaster*:					
Gene mutations	2–7 weeks	moderate	moderate	good to excellent	
Chromosome aberrations	weeks	moderate	moderate		good to excellent
Mammals:					
Dominant lethal mutations	2–4 months	moderate to high	moderate		unknown
Translocations	5–7 months	moderate to high	moderate		potentially very good
Blood or bone marrow cytogenetics	1–5 weeks	moderate	moderate		potentially good
Specific locus mutations	2–3 months	high to very high	high to very high	unknown	

Source: John W. Drake et al. 1975. "Environmental mutagenic hazards." *Science* 187 (February 14): 507.

drifted into obsolescence, stand as the material embodiment of research activism. They demonstrate, perhaps more clearly than anything else the scientist-activists' movement created, how science and activism were fused together and institutionalized into a mode of scientific practice that became one of the defining features of genetic toxicology.

Conclusion

Concentrated outside the university system and in the absence of either a ready-made labor force or market-creating legislation, EMS members in the United States essentially built the institutional foundations of genetic toxicology from the ground up. They did so collectively by creating mechanisms for recruiting and training scientists, coordinating research, standardizing research tools and practices, and undertaking public outreach and education. Maintaining a strict distance from groups with clear-cut political and/or economic interests represented an organizational strategy for balancing the competing claims that research promoted by the EMS was at once socially relevant and unblemished by social bias. Policing that boundary, in turn, enhanced the legitimacy of genetic toxicology as well as advancing the EMS's own influence and autonomy. Keeping environmental politics ostensibly out of genetic toxicology was ultimately a means for gaining and holding on to that authority.

Policing the science/politics boundary was also, paradoxically, a strategy for accomplishing the EMS's own political work. At least for a time, the boundary between environmental health research and activism blurred *within* the EMS. As the organization's commitments to the production of accurate knowledge, to the integrity of the new interdiscipline, and to the reduction of environmental genetic hazards converged, collecting and disseminating mutagenicity data, on the one hand, and building the political and moral case against the indiscriminate use of mutagenic chemicals, on the other, came to be treated as complementary and mutually reinforcing projects.

The EMS was a central player in the campaign to institute the new order of environmental inquiry that, by 1976, genetic toxicology had come to represent (Drake et al. 1975). Ironically, the new field's rapid rise may be traced in part to the EMS's effectiveness at maintaining a publicly visible boundary between environmental science and environmental politics while simultaneously subverting that same boundary within its own organizational domain.

7

Conclusion

Environmental Knowledge Politics in Practice

> The bottom line is human germinal mutation and the translation of this into effects on human welfare. We still have no reliable way to move from DNA damage, however precisely measured, to human well-being n generations from now.... This does not mean that the new cellular and molecular information and ever more precise testing systems are not enormously useful. Lowering the mutation rate or preventing its increase is good, even if we don't know how good.
>
> —James F. Crow, "Concern for Environmental Mutagens"

In less than a decade, the chemical mutagens that geneticists had once used exclusively as tools in experimental research gained new meaning as environmental problems, and a new interdiscipline emerged to claim "environmental mutagenesis" as its central topic. These transformations changed the way genetics knowledge was made and who made it. They also changed how environmental health specialists and policy makers interpreted the human consequences of chemical pollution. What accounts for the rise of genetic toxicology?

Here we should recall Charlotte Auerbach's (1978:183) remark that, during the 1960s, she did not feel "inspired" to test individual chemicals for mutagenicity because doing so would have meant "testing the hypotheses of chemists, and where would be the fun of that?" Genetic toxicology required more than facts or institutional space to grow. It required an organizational incentive to "inspire" boundary crossing. What mutation researchers lacked in the 1960s was a shared sense of urgency and social responsibility to put issues of public health and safety above disciplinary conceits. The missing ingredient, I have argued, was a scientists' social movement—a collective and sustained effort by scientists to reshape the structures of science in order to more effectively respond to social and environmental concerns.

Taking a page from Rachel Carson's *Silent Spring*, scientist-activists described the human gene pool as a uniquely fragile natural resource threatened by the unintended consequences of modern industrial practices. The rhetorical case for genetic toxicology melded the evolutionary concerns well known to geneticists with the preservationist spirit of a new environmentalism to create a powerful symbolic package with broad appeal and urgency. At the center of these appeals, the EMS played a critical role in organizing collective action to coordinate policy work and redesign training curricula and laboratory research. The permeable boundaries that came to mark genetic toxicology as an interdiscipline are the result of scientist-activists' collective attempts to regulate the mix of science and politics flowing into and through mutation research.

Ironically, social movements are hobbled by their own success (Tarrow 1998). The Toxic Substances Control Act (TSCA), which created a formal market for genetic toxicology knowledge, signaled the beginning of the end. Two years later the EPA released standardized protocols for mutagenicity testing (Prival and Dellarco 1989). With regulatory reforms in place, and with the standardization of mutagenicity bioassays, genetic toxicology testing became business-as-usual for a growing number of emerging private laboratories. By the late 1970s, the tide of scientist activism had begun to ebb, and the movement phase of genetic toxicology's historical development largely subsided.

Measuring Success in Genetic Toxicology

I have argued that the movement to create genetic toxicology was highly successful in its initial bid to reorganize disciplinary boundaries and to infuse environmental values into the logic of experimental design in mutation research. One handy metric for assessing these claims are the four action recommendations urged by the NIH Genetics Study Section in 1966 and summarized in James Crow's (1968) article "Chemical Risk to Future Generations." These scientists called for the creation of a chemical mutagenicity data registry, a formal chemical mutagenicity screening program, more sensitive and cost-effective bioassays, and a program for genetically monitoring human populations. By this yardstick, the movement's short-term achievements were impressive. Scientists at Oak Ridge began developing the EMIC database in 1970, and mutagenicity testing for new chemicals became the law of the land in 1976 (under TSCA), which in turn helped fuel interest in the development of dozens of short-term bioassays over the next ten years for testing the mutagenic potential of chemicals across a range of experimental organisms and end points. Only the last recommendation to move forward with human population monitoring, an effort that Crow (1968:117) admitted would be "both difficult and expensive," fell short of expectations. Today, there is only one mammalian germ-cell assay used for population monitoring efforts.[1]

Institutional Stability and Change

We can also take a longer-term view on the issue. With hindsight, it seems clear that many of the institutions that scientist-activists created thirty-some years ago remain central to communication, education, and research in genetic toxicology—even as the field has grown to incorporate new organizations, curricula, and practices far beyond the reach of those incipient structures. The EMS, for example, remains a vital and energized professional society. It is a relatively small but stable organization whose membership levels have hovered around the 1,000 mark for the better part of two decades and whose members have spun off three associated regionally based organizations.[2] Since 1983, the EMS has published *Environmental and Molecular Mutagenesis,* a journal of original research that is, like its parent organization, small but well respected within the broader mutation research and environmental health communities. Over the years, EMS members have worked to build a network of thirty-four national, pan-national, and international professional societies that promote communication and research on environmental mutagenesis and related topics.[3] If the EMS has maintained a narrow focus on environmental mutagenesis (perhaps, in the view of some, too narrow), genetic toxicology is represented in many other larger professional societies, such as the Society of Toxicology and the American Association of Cancer Research.[4]

A long-standing focus of the EMS and its related societies has been the organization of international training workshops in environmental mutagenesis and genotoxicity. These workshops, many of which are specifically designed to provide practical skills training to environmental health specialists in developing countries, are a direct legacy of programs begun under Alexander Hollaender's leadership in the early 1970s. In the United States, genetic toxicology training has entered core curricula in environmental and public health sciences. While regular departments of genetic toxicology still do not exist in U.S. medical schools or universities, the interdiscipline has emerged as a standard feature of public health and environmental science programs. It is also now characteristic to find faculty in the life sciences (broadly defined) who conduct research on various aspects of biological effects research relating to genetic toxicology and who train their students in those methods. One way or another, universities and medical schools now are producing genetic toxicologists.[5]

If genetic toxicology research has expanded and diversified over the past three decades, it has also stayed at home in the federal science system where it originated. The Oak Ridge National Laboratory Biology Division is no longer the powerhouse it once was for chemical and radiation mutagenesis. When I visited there in 1997, the famed "mouse house"—once the largest mammalian genetics laboratory in the world—had fallen on leaner times, with its financial base and its animal population considerably diminished. A new laboratory, slated to open in 2004, is expected to put mammalian genetics at Oak Ridge back on the map.[6] In other corners of federal science, genetic toxicology thrives in laboratories like the NIEHS's National Center for Toxicogenomics (NCT). Created in 2000, the NCT is an important component of a federal initiative that will enfold genetic toxicology within the rapidly emerging bioinformatics-driven arenas of genomics and protenomics (National Center for Toxicogenomics 2003). With a well-funded, high-visibility research center organized around cutting-edge technologies and computational power that promoters promise will revolutionize environmental health science, genetic toxicology has formally entered the genomic age (Shostak 2003a).

In this reinventive context, striking parallels remain between the NCT's promotion of toxicogenomics and the EMS's earlier program for building genetic toxicology. Recent calls for collaborative research linking scientists from private industry, academia, and other government laboratories and a commitment to balancing "discovery science" with "hypothesis-driven science" ring especially familiar (National Center for Toxicogenomics 2003:5). The use of genetic technology as a mechanism for interdisciplinary interaction also links the old and new genetic toxicology in form, if not in content. In the early 1970s bioassays functioned as "boundary objects" that tied public health specialists to environmental mutagenesis (Star and Griesemer 1989).[7] Today, environmental health sciences are linked to molecular biology and genomics via "high sensitivity, rapid throughput technologies" that are being designed to allow researchers to "identify toxic substances in the environment and those populations at the greatest risk of environmental diseases" (National Center for Toxicogenomics 2003:2, iii). If interdisciplinarity has intensified under this high-tech scenario, the fundamental dynamics underlying those interactions continue to closely parallel the past.

Today, of course, talk about hybrid knowledge and collaborations that span public- and private-sector science is as commonplace among those who conduct, administer, and finance research as it is among those who study these interactions. But the comparison serves as a reminder that, in 1970, the arc between industry and university laboratories was wider and often more harrowing to cross. For better and for worse, the scientists' movement to establish genetic toxicology was, if nothing else, prescient in inviting industry into the EMS and in encouraging research biologists to take seriously the public health implications of their laboratory bench work.

Environmental Values

Still another gauge for assessing the movement's success is whether the environmental values that initially transformed the logic of bioassay design in chemical mutagenesis have continued to influence the trajectory of research on environmental mutagenesis. Since the field's formation, genetic toxicologists have approached the problem of environmental mutagenesis through two basic strategies that gained popularity sequentially, both of which have informed regulatory policy and risk assessment (Samson 2003). The first, which dominated genetic toxicology research through the 1980s, emphasized hazard identification. During this period, a dynamic genetic toxicology testing economy emerged, and state-sponsored review panels evaluated the mutagenicity literature and summarized the cumulative findings in published research reports.[8] The second strategy, which became dominant during the 1990s and remains so today, emphasizes the molecular mechanisms of mutation. Spurred in part by technological advances in molecular imaging, DNA replication, and transgenic systems, scientists have gained a much more complex and nuanced understanding of the genotoxic effects of chemical agents and how that knowledge translates into quantitative risk assessment (Preston and Hoffman 2001). In each of these approaches, the enduring role of environmental values in shaping genetic toxicology seems to be apparent.

But there is also room for a more critical interpretation. As Daniel Sarewitz (1996) points out in discussing what he calls "the myth of infinite progress," more knowledge does not automatically produce social or environmental benefits. Instead, he argues, the notion that "more is better" is often used as a legitimating ideology by those with vested interests in growing the R&D economy, whether or not the resulting knowledge leads to healthier communities and environments. If the infusion of environmental values into scientific discourse on mutation research once directed concerned attention to the long-term impacts of environmental mutagens on human populations, there is less evidence that those arguments retained their rhetorical influence as genetic toxicology institutions and practices stabilized.

Although the scientists' movement initially drew moral, political, and scientific legitimacy from the urgent need to prevent what James Crow called a "genetic emergency" (1968:113), the intensity of geneticists' concern over the long-term implications of chemical germ-cell mutagenesis has been gradually but steadily nudged aside by other concerns. One of these concerns has been cancer. In the decade or so following the genetic toxicology movement's early institutionalizing successes, its discursive center of gravity shifted to somatic-cell mutations in the wake of Berkeley biochemist Bruce Ames's (Ames et al. 1973) provocative claim that "carcinogens are mutagens." While the predictive value of mutagenesis was initially overplayed, the fast, simple, and cost-effective short-term tests for mutagenicity that Ames and others developed have been enshrined in regulatory requirements and in biomedical research more generally as carcinogenicity screens.[9] Environmental mutagenesis has gained lasting importance not as an environmental science in its own right but as the field that carries the canary into the coal mine of cancer science.

Mutagenesis's subordination to carcinogenesis was not a new one. As H. J.

Muller complained to Joshua Lederberg in 1950, in a letter that is uncanny in its predictive accuracy,

> My own work is dependent on cancer grants and, as I expected, the cancer people are pulling the purse strings tighter when it comes to giving money for genetics. It is not right that mutation work should have to be a tail to the cancer kite. I think the time has come when it ought to be recognized in its own right and that we ought to make an effort to get a movement to support it started by the NRC, unless some more suitable agency can be found. (Lederberg 1997:7)

The more recent canonization of carcinogenicity as the biological end point of central significance in genetic toxicology research is in some ways emblematic of a more general trend away from the evolutionary and environmental uncertainties that drove Muller's and Lederberg's earlier concerns.

In broad terms, a series of subtle and not-so-subtle shifts in emphasis seem to have weakened genetic toxicology's initial environmentalist thrust. Although there are a number of ways one may parse the issue, dominant trends in contemporary genetic toxicology, while by no means definitive or linear, seem to move in a common ideological direction: from germ-cell to somatic-cell damage; from populations to individuals; from future generations to those now living; from ultimate to proximate causes. For example, there is little talk today about mutational loads in populations but loads of talk about genetic therapies to treat individuals. Scientists' early concerns about unintended exposures to mutagens in pollution, pesticides, and food additives have taken a backseat to concerns about prescribed exposures from pharmaceutical drugs. Where scientist-activists once decried the federal government's lack of regulation of potential genetic hazards, today the NIEHS is at least as concerned about preventing "needless and expensive over-regulation" (National Center for Toxicogenomics 2003:6).

The consequences for environmental health are real enough, even if we have only the unsettling absences of knowledge to use as a meter. Uncertainty reigns in environmental effects research, no less in genetic toxicology than in climate science. Although three decades have now passed, "very few of the published studies of cytogenetic population monitoring for individuals have analyzed the appropriate endpoint for detecting the genetic effects of long-term exposure to chemicals" (Preston and Hoffman 2001:344). While the lack of credible longitudinal biomonitoring data can be explained away by methodological difficulties involving long lists of potential "confounding factors" in selecting experimental and control groups, it is also clear that preserving the genetic integrity of future generations has not been a research priority for most genetic toxicologists or for those who finance their work.[10] The problem of extrapolating results from mutagenicity tests in submammalian systems to human beings, as the epigraph for this chapter suggests, also remains vexing. NCT promoters see the new genomic sciences' focus on gene-environment interaction as the way to finally solve this problem and increase the predictive value of genotoxicity risk assessments. Others caution that while genomic technologies hold "great promise for establishing a cell's response to exposure to chemical or physical agents in the context of normal cellular patterns of gene expression, it remains to be established how to analyze the vast amounts of data that can and are being obtained and what magnitude of change in gene expression constitutes an adverse effect. . . . Extrapolating the responses to organs and whole animals represents a challenge still to be addressed" (Preston and Hoffman 2001:345).

Rethinking Science Activism

The rise of genetic toxicology and the retrenchment of the environmentalist impulses that first ignited it provide an opportunity to reconsider the social significance of activism in environmental science. The scientists' movement that established genetic toxicology is sociologically interesting precisely because scientist collective action was so unlike the more familiar forms of radical science activism emanating from the socialist movements of the 1920s and 1930s, the student movements of the 1960s and 1970s, or the antinuclear movements of the 1980s. The movement to create genetic toxicology was not led by students politicized by the war in Vietnam or by Earth Day protests but by seasoned professionals—geneticists mostly, along with a few biochemists, toxicologists, and pharmacologists whose reputations and careers were firmly established. These scientists did not form unions or march with workers (Kuznick 1987; McGucken 1984). They did not all oppose the war in Southeast Asia, nor were they particularly inclined toward the more radical views promoted by domestic environmentalists.[11] They certainly did not disrupt professional conferences or demand fundamental changes in scientists' relationship to machineries of state-sponsored violence (Moore forthcoming). Their goals were more modest and their actions far less confrontational.

This was a reform movement. Scientists argued for the importance of putting existing knowledge to new uses, for identifying new problems by prioritizing public health over basic research, and for reorganizing the disciplinary boundaries of science to accommodate research on complex environmental problems whose solutions required multidisciplinary approaches. These modest goals had deceptively profound impacts. The movement helped change the way knowledge was produced, by whom, and for what purposes. It created the legal, organizational, and technological foundations for a new public-service genetics.

Although these scientists did not take their protest to the streets, and even though many of the people I have referred to throughout this study as "scientist-activists" would probably not recognize themselves in that label, the movement to establish genetic toxicology clearly embodied and expressed a politics of environmental knowledge. Their collective actions in the name of environmental health demonstrate that there are less visible and less direct ways that scientists' movements can transform scientific knowledge, practices, and institutions. Too often, discussions of the relationship between science and social movements are based on undertheorized perceptions of the two as organizationally and epistemologically distinct phenomena. It is only recently that science studies scholars have begun to examine systematically how social movements and science interconnect, and to date there are very few studies that take scientist activism itself as a topic for serious analysis.

Common images of environmental science activism are those popularizers of environmental ideas that Rae Goodell (1977) has called "the visible scientists." Barry Commoner, Paul Ehrlich, E. O. Wilson, and Steven Schneider are among those relatively few scientists who have played important roles in translating the technical details of ecological disruption into the metaphors that the media carries into contemporary popular culture. They have brought environmental crisis, the population bomb, biodiversity loss, and global warming (to name but a few) onto the public square (Hannigan 1995; Mazur and Lee 1993). That most environmental scientists are not culturally visible as icons of environmentalism does not mean that environmental advocacy of one sort or another in science is uncommon; recent reports suggest just the opposite (Brown 2000).

For their part, genetic toxicology activists did most of their organizational and promotional work within their scientific communities and networks and, as often as not, inside their own laboratories and in classrooms, where it crucially mattered. Their actions may have presaged by twenty years the dominant environmental slogan of the 1990s to "think globally, act locally," but that in essence describes the movement's general approach to reshaping the form and content of mutation research. Among other things, this undercover approach to environmental politics suggests that science activism is not ephemeral to scientific practice. Rather, it reinforces science studies scholars' contention that environmental knowledge politics and practices are mutually constituted.

It also demonstrates that scientist activism can take many unexpected forms. There is a sense in which activism in genetic toxicology is counterintuitive because much of what these scientists actually did was in many ways identical to what most scientists typically do in professional life: they tinkered, puzzled, innovated, shared and promoted their ideas, and competed with one another for grants and status. Students of knowledge politics impose unnecessary constraints on analyses that begin with rigid assumptions about what kinds of social actions properly count as science activism. This study demonstrates that under specific conditions, normative actions can generate transformative change in science. Here, context is critical. As in social protest more generally, scientist-activists tailor strategies, tactics, and modes of collective action in relation to the structural conditions of their work. In the case of genetic toxicology, conventional behavior motivated by environmental values bore contentious outcomes. One of those outcomes was a new interdiscipline. Another was the politicization of scientists' identities.

Collective identities are constructed through collective action; regular folks are politicized in the heat of protest (Morris and Mueller 1992). Following their participation in the movement to build genetic toxicology, at least some of the core activists have gained lasting reputations as political activists of one sort or another. Sam Epstein, for example, has devoted much of his time and energy during the last thirty years waging war on the "war on cancer." He has been a vocal critic of the cancer research establishment's prioritization of mechanisms research and the private interests of the pharmaceutical industry over public health and disease prevention (Epstein 1979; Epstein et al. 1982; see also Proctor 1995:ch. 3).[12] Epstein's continuous attacks on the National Cancer Institute, the American Cancer Society, and the pharmaceutical drug industry have won him ire in government and industry circles, but his ongoing efforts to link environmental pollution to cancer rates have tendered awards from the National Wildlife Federation, the Citizen's Clearinghouse for Hazardous Waste, and the National Coalition Against the Misuse of Pesticides, in addition to various awards from universities and scientific and medical societies.

If Epstein's strategy for fomenting change in health science has been to publicly confront the major institutional actors in an effort to expose the underlying economic interests that shape research agendas and knowledge outcomes in environmental health sciences, former EMS treasurer Marvin Legator has taken his political convictions onto the streets. Legator is well known among environmental justice groups along the Texas and Louisiana Gulf Coast.[13] After leaving the FDA in the early 1970s and following a short stint at Brown University, Legator helped establish an environmental toxicology program at the University of Texas Medical Branch in Galveston. For years, Legator has worked with local communities, helping them document evidence of health impacts from toxic exposures. He has conducted community health surveys in minority and working-class communities bordering petrochemical plants and oil refineries and has served as an expert witness in legal suits against industrial polluters (e.g., Morris 1997). Legator has also trained community activists to design, collect, and analyze health survey data and to collect air, soil, and water samples to document exposure to chemical hazards (Legator et al. 1985).

Other core activists have pursued less radical tactics for putting their knowledge and skills as scientists to political use. Bruce Ames was for many years at the forefront of efforts to publicize the potential hazards of mutagens in food and consumer products. Since the 1990s, Ames has become an equally vocal antienvironmentalist, at times lining up opposite Sam Epstein in debates conducted through science journals. Much of Ames's more recent research has focused on naturally occurring mutagens, research which he uses to downplay the role of environmental chemicals in causing cancer (Proctor 1995:ch. 6). Although Ames's environmental message is a reactionary one, his tactics, by and large, are not. Another rather more conventional approach is illustrated by Matthew Meselson's longtime efforts to shape chemical and biological weapons (CBW) policy. Described recently as an activist who works mostly "behind the scenes" (Brickley 2002), Meselson codirects the Harvard-Sussex Program on CBW Armament and Arms Limitation, a program that seeks "to promote the global elimination of CBW weapons and to strengthen the constraints against hostile use of biomedical technologies."[14] He has investigated releases of anthrax in the Soviet Union during the 1980s (Guillemin 1999) and in the U.S. postal system since 2001.

The political biographies of these four former genetic toxicology scientist-activists describe different points along intersecting continua of scientist activism that range from conventional to confrontational and from radical to reactionary. The institutions they target and the impacts of their tactics also differ. Importantly, however, none has been forced to sacrifice scientific credentials for political convictions.

Others have continued to influence environmental health and science policy through traditional advisory channels. James Crow, for example, has been a regular member on National Academy of Sciences advisory committees shaping federal policy on low-level radiation exposures, nuclear and alternative energy, chemical environmental mutagens, and DNA technology in forensics.[15] James Neel supplied congressional testimony in support of the International Biological Program (in 1970), criticizing Department of Defense assumptions on the extent of effects of limited nuclear war (in 1975), and against a House bill providing that life begins at conception (in 1981).[16]

For still others, the values and social networks generated and nurtured circa 1969 continue to shape modest and diverse efforts to build from the legacy of the genetic toxicology movement. Many examples emerged from my conversations with EMS members. One scientist admitted to channeling funds from grants for mainstream mutagenesis research "under the table" for environmental research he could not otherwise finance. Another has devoted much of his career promoting simple-to-use plant bioassays as biomonitors of environmental stressors in grassroots education programs in developing countries (Ma 1995). Others have argued for extending genetic toxicology tools and assessment practices from human health to ecological health.[17] Efforts by genetic toxicologists at the EPA have recently begun to re-ignite interest in human germ-cell mutations and population risk.[18]

Of course, not all of the people involved in the early development of genetic toxicology continued their enthusiasm for this new approach to genetics research or its representative organization. Several scientists I interviewed expressed discontent with the early emphasis on hazard identification as not constituting "real science." Others complained that the EMS did not grow in parallel with the expansion of genetic toxicology because the organization was slow to embrace molecular biology and because leaders resisted calls to shed the "environmental" label from the organization's name. This study suggests that such complaints are misplaced. Drawing sharp dichotomies that separate science from politics, basic research from applied, or research from activism misses the point. Molecular biology is no less political for its attention to questions about how mutagenesis and carcinogenesis occur. And environmental mutagenesis is no less scientific for linking mutational phenomena to broader questions about human exposure to chemical agents and the health of future generations. Sylvia Tesh (2000) has recently described the continuing need for an "environmentalist" science—one that can accommodate the tension between facts and values. An environmentalist genetic toxicology is one that seriously attends to basic questions of biology but also maintains an activist orientation toward environmental health research. Such a science will recognize that answers to the questions that molecular biology asks are not in themselves solutions to environmental problems and that conflating the two is often a recipe for political inaction—an environmental hazard of a different sort.

Conclusion

The environmental writer Bill McKibben (1990:90) has argued that we live in a "post-natural" world in which all nature now contains a social imprint. If he is correct, synthetic chemicals are the main reason why. At the dawn of the twenty-first century, human-made chemicals are globally ubiquitous in the products that industry creates, markets, and sells and that consumers the world over buy, use, and throw away. Chemicals are "always everywhere," constantly intermeshing with our daily lives. A curious aspect of our postnatural condition is that synthetic chemicals are also, in important respects, nowhere. As material entities, chemicals themselves are largely invisible. We do not see the substances that preserve our food, green our lawns, whiten our clothes, or thin atmospheric ozone. Unless one works in a factory that produces or uses chemicals or lives near places where chemicals are stored or dumped as waste, most people, most of the time, tend not to notice our chemical environment outside the polemic abstractions of industrialists who celebrate, and environmentalists who vilify, "the chemical age." We take the things themselves—as they exist now—largely for granted.

The future impacts of those same chemicals on individual bodies, on communities, and on ecosystems come wrapped in still more layers of abstraction. Concern for "the children of our grandchildren" and "ecological sustainability" may deepen fears about the unintended consequences of our increasingly chemical lives, but the moralistic rhetoric of global environmentalism also ensures that political action that addresses those fears will have to rely on answers to questions that are essentially unknowable. How do we in the twenty-first century govern chemical effects in human populations of the twenty-second? This question, among others, shapes ongoing debate about society's relationship to nature and to the synthetic chemicals that now infuse nature, including us. The scientist-activists who created genetic toxicology transformed the institutions that made and ordered environmental knowledge through collective action that was vigilant, coordinated, and sustained. Today, confronting the chemical consequences of a postnatural world may require much more but certainly demands nothing less.

APPENDIX A

Scientists Interviewed

Name	Employer/Position at Time of Interview	Interview Date
Seymour Abrahamson	Department of Zoology (Emeritus), University of Wisconsin–Madison	Apr. 15, 1997
Bruce Ames	Department of Molecular and Cell Biology, University of California–Berkeley	Nov. 13, 1997
Herman Brockman	Department of Biology (Emeritus), Illinois State University	Nov. 7, 1997
David Brusick	Covance Laboratories, Vice President, Toxicology Division	Apr. 18, 1998
Larry Claxton	National Health and Environmental Effects Lab, EPA	Mar. 23, 1998
Donald G. Crosby	Department of Environmental Toxicology, University of California–Davis	Mar. 6, 1998
James F. Crow	Department of Genetics (Emeritus), University of Wisconsin–Madison	Apr. 18, 1997
Frederick J. de Serres	Technology Planning and Management Corp. (Research Consultant)	Apr. 20, 1997
John W. Drake	Laboratory of Molecular Genetics (Chief), NIEHS	Mar. 1, 1998
Samuel S. Epstein	School of Public Health, University of Illinois Medical Center	July 14, 1997
James M. Gentile	New Hope College (Dean of Biology)	Mar. 25, 1997
James W. Gillett	Superfund Basic Research and Education Program (Director), Cornell University	June 18, 1997
William F. Grant	Department of Plant Science, McGill University	Apr. 21, 1997
Philip Hartman	Department of Biology (Emeritus), Johns Hopkins University	Dec. 22, 1997

Name	Employer/Position at Time of Interview	Interview Date
David S. Hinton	Environmental Toxicology Program, University of California–Riverside	Mar. 31, 1998
Eugene Kenega	Dow Chemical Corporation (retired)	Oct. 19, 1998
Te-Hsiu Ma	Department of Biological Sciences, Western Illinois University	Mar. 24, 1998
Heinrich Malling	Laboratory of Molecular Genetics, NIEHS	Apr. 22, 1997
Mortimer Mendelson	Biology and Biotechnology Research Program, Lawrence Livermore Laboratory	Mar. 25, 1998
Kristien Mortelmans	Stanford Research Institute, International	Mar. 25, 1998
James V. Neel	Department of Human Genetics (Emeritus), University of Michigan	Apr. 22, 1997
Tong-man Ong	Toxicology and Molecular Biology Branch, Health Evaluation Research Division, NIOSH	Apr. 18, 1998
Michael J. Plewa	Department of Crop Sciences, University of Illinois	Mar. 24, 1998
R. Julian Preston	Chemical Industry Institute of Toxicology	Sept. 29, 1997
Liane B. Russell	Oak Ridge National Laboratory Biology Division	Sept. 26, 1997
Michael Shelby	Laboratory of Toxicology, Environmental Toxicology Program, NIEHS	Oct. 2, 1997
Raymond W. Tennant	Laboratory of Environmental Carcinogenesis and Mutagenesis (Chief), NIEHS	Oct. 2, 1997
Larry Valcovic	U.S. EPA	Apr. 21, 1997
John Wassom	Human Genome and Toxicology Group, Oak Ridge National Laboratory	Apr. 21, 1997; Sept. 25, 1997
Michael Waters	National Health and Environmental Effects Lab, EPA	Sept. 29, 1997
Errol Zeiger	Environmental Toxicology Program, NIEHS	Oct. 1, 1997

APPENDIX B

Institutionalizing Events in Environmental Mutagenesis/ Genetic Toxicology, 1964–1976

Year	Event	Place
1964	Symposium, "Molecular Action of Mutagenic and Carcinogenic Agents" (sponsored by ORNLBD)	Gatlinburg, Tenn.
1965	Brookhaven Symposium in Biology, no. 18, "Genetic Control of Differentiation"	Upton, N.Y.
1966	Genetics Study Section (NAS) conference on chemical mutagens	Woods Hole, Mass.
1967	Panel discussion on "radiation and chemical mutagenesis" at Habrobracon-Mormoniella Conference (parasitic wasps)	Oak Ridge Biology Division, Oak Ridge, Tenn.
	Conference on "mutation research on microorganisms"	Edinburgh, Scotland
1968	Genetics Study Section report published in *Scientist and Citizen*	
	International symposium, "Genetic Effects of Radiation and Radiomimetic Chemicals"	Kyoto, Japan
	"Roundtable on Mutagenesis"	Gaithersburg, Md.
1969	Environmental Mutagen Society (EMS) formed	Washington, D.C.
	Ciba Foundation symposium, "Mutation as a Cellular Process"	London, England
	Panel, "Interactions between the Genetic Apparatus and Exogenous Agents," 23rd Annual Symposium on Fundamental Cancer Research (Dartmouth College)	Hanover, N.H.
	1st issue of *EMS Newsletter* published	Oak Ridge, Tenn.
	EMIC formally established	Oak Ridge, Tenn.
	Symposium,"Drugs of Abuse: Their Genetic and	San Francisco, Calif.

Year	Event	Place
	Other Chronic Nonpsychiatric Hazards" (cosponsored by the Center for Studies of Narcotic and Drug Abuse, NIMH, and EMS)	
	Environmental mutagenesis program established at ORNLBD (de Serres coordinates)	Oak Ridge, Tenn.
	Roundtable discussion on environmental mutagens at Genetics Society annual meeting	Madison, Wis.
	Symposium, "Chemical Mutagenesis in Mammals"	Mainz, W. Germany
	Inauguration of the Central Laboratory for Mutagenicity Testing	Freiburg, Germany
	Mutagenesis workshop (NIEHS)	Bethesda, Md.
1970	1st annual meeting of U.S. EMS	Washington, D.C.
	Meetings between EMS and National Academy of Sciences members	Washington, D.C.
	Informal conference, "Repair and Mutation in Microorganisms" (Pisan Lab. for Mutagenesis and Differentiation), 23 papers presented	Pisa, Italy
	Symposium, "Fundamentals of Mutagenicity Testing"	Woods Hole, Mass.
	Symposium, "Mammalian Radiation Genetics"	Neuherberg, Germany
	European Environmental Mutagen Society (EEMS) formed	Neuherberg, Germany
	Molecular Basis of Mutation published (Drake)	
	Chemical Mutagenesis in Mammals and Man published (Vogel and Röhrborn, eds.)	
	Chemical Mutagens in Man's Environment published (Fishbein et al., eds.)	
	Cell Mutation Unit established at University of Sussex	Brighton, Sussex, Great Britain
	Conference, "Evaluating Mutagenicity of Drugs and Other Chemical Agents" (Drug Research Board)	Washington, D.C.
	International symposium, "Chemical Mutagenesis as a Problem in Medicine" (Ciba Geigy)	Basle, Switzerland
1971	1st annual meeting of EEMS	Leiden, Netherlands
	Gesellschaft fur Strahlen- und Umweltmutationsforschung (GUM) (Environmental Mutagen Research Society) formed	Fed. Rep. of Germany
	"Workshop on Mutagenicity," Brown University	Providence, R.I.

Year	Event	Place
	(EMS, FDA, NIEHS, Pharmaceutical Manufacturers Association, Drug Research Board)	
	Roundtable session, "Genetic Hazards from the Environment," Fourth International Congress of Human Genetics	Paris, France
	Gruppo Italiano Mutageni Ambientali (GIMA) organized as a national section of the EEMS	Pisa, Italy
	Chemical Mutagens: Principles and Methods for Their Detection, vols. 1–2, published (Hollaender, ed.)	
1972	NIEHS Mutagenesis Branch organized	Research Triangle Park, N.C.
	Joint Industry–Government–University Study on Dominant Lethal Cytogenetic Methods completed	
	5-week training course on "chemical mutagenesis" in Latin America (Organization of American States)	
	Conference, "Detection of Somatic Mutations in Humans"	
	Fogarty Center workshop, "Mutagenic Contaminants in the Environment"	
	3rd annual EMS meeting; includes a symposium on "new techniques in mutagenicity testing" and papers on techniques for evaluating mutagenicity of atomic bomb survivors and of industrial mutagenic hazards	Cherry Hill, N.J.
	Symposium, "Mutagenic Hazards from the Environment," Medical and Biological Federation meetings	Amsterdam, Holland
	2nd annual EEMS meeting	Pilsen, Czech.
	Symposium, "Chemical Mutagenesis in Microorganisms and Plants"	Ibaraki-ken, Japan
	Japanese Environmental Mutagen Society (JEMS) formed and 1st annual JEMS meeting held	Mishima, Japan
	Environmental Mutagenicity Workshop; demonstrations to members of British Chemical and Pharmaceutical Industry (University of Edinburgh)	Edinburgh, Scotland
	International Workshop on Mutagenicity Testing of Drugs and Other Chemicals	Zurich, Switzerland

Year	Event	Place
	Last issue of *EMS Newsletter* published (no. 6)	Oak Ridge, Tenn.
	GUM joins EEMS as a national section of W. Germany	
	Section on Environmental Mutagenesis of the Czechoslovak Biological Society formed as a national section of the EEMS	Prague, Czech.
1973	"Environmental Mutagenesis and Related Subjects," a new section of *Mutation Research*, begins publication (de Serres, ed.)	Oak Ridge, Tenn.
	The Testing of Chemicals for Carcinogenicity, Mutagenicity, and Teratogenicity published by the Ministry of National Health and Welfare (Canada)	
	Workshop, "Problems and Perspectives in Mutation Research" (org'd. by Sobels)	Noordwijkerhout, Netherlands
	3rd annual EEMS meeting; Symposium, "Caffeine as an Environmental Mutagen and the Problem of Synergistic Effects"	Uppsala, Sweden
	1st International Conference on Environmental Mutagenesis	Asilomar, Calif.
	International Association of Environmental Mutagen Societies formed	Asilomar, Calif.
	2nd annual JEMS meeting	Mishima, Japan
	Conference, "Dose and Effect of Mutagenic Chemicals"	Bad Kruzingen, Germany
	International symposium, "Testing Mutagenic Effects of Environmental Contaminants"	Prague, Czech.
	"1971 survey of literature reporting chemical mutagenesis" (EMIC report)	
	"The mutagenicity and teratogenicity of a selected number of food additives" (EMIC report)	
	"Panel on environmental mutagenesis and carcinogenesis" formed (U.S.-Japan Cooperative Medical Science Program)	
	"Panel on mutagenesis" formed (U.S.-USSR agreement on cooperation in the field of environmental protection)	
	Interagency Panel on Environmental Mutagenesis (U.S. Department of Health, Education, and Welfare) formed	

Year	Event	Place
	Workshop, "Evaluation of Mutagenicity Data and Extrapolation to Man"	
	Workshop, "DNA Repair Systems in Fungi"	
	Chemical Mutagens: Principles and Methods for Their Detection, vol. 3, published (Hollaender, ed.)	
	"Molecular and environmental aspects of mutagenesis," 6th Rochester International Conferences on Environmental Toxicity (University of Rochester)	New York, N.Y.
	"Chemical mutagenesis in laboratory mammals: A bibliography on the effects of chemicals on germ cells" (EMIC report)	Oak Ridge, Tenn.
	"The mutagenicity and teratogenicity of a selected number of food additives: EMIC/GRAS literature review" (EMIC report)	Oak Ridge, Tenn.
	"Chemical mutagenesis, a survey of the 1971 literature" (EMIC report)	Oak Ridge, Tenn.
1974	4th annual EMS meeting; "Mutagenesis" symposium, held jointly with Society of Toxicology	
	4th annual EEMS meeting	Heidelberg, Germany
	3rd annual JEMS meeting	Tokyo, Japan
	Workshop, "Mutagenicity of Chemical Carcinogens"	Honolulu, Hawaii
	Conference, "Monitoring the Problem of Mutagenesis and Carcinogenesis in Man"	Tokyo, Japan
	Workshop, "Long-term Toxicity of Antischistosomal Drugs"	
	Workshop, "Mechanisms of Chemical Carcinogenesis"	
	"Chemical mutagenesis: a survey of the 1972 literature" (EMIC report)	Oak Ridge, Tenn.
1975	"Reviews in Genetic Toxicology" begins as special section of *Mutation Research*	
	"Committee 17" report published in *Science*	
	Handbook of Mutation Testing Procedures (Legator et al., eds.) begins publication "article-wise" in *Mutation Research*	
	Workshop, "Basic and Practical Approaches to Environmental Mutagenesis and Carcinogenesis"	New York, N.Y.

Year	Event	Place
	Workshop, "Chemical Mutagens in the Netherlands"	Leiden, Netherlands
	6th annual EMS meeting	Miami Beach, Fla.
	Workshop, "International Coordination of Environmental and Chemical Mutagenesis Studies"	Miami Beach, Fla.
	International Symposium, "Genetic Hazards to Man from Environmental Chemicals" (York University)	Toronto, Canada
	Conference, "In Vitro Mutagenicity and Carcinogenicity Tests"	Seattle, Wash.
	Department of Radiation Genetics (State Univ. Leiden) changed to Department of Radiation Genetics and Chemical Mutagenesis	Leiden, Netherlands
	4th annual JEMS meeting	Kyoto, Japan
	Section on Environmental Mutagens formed within the Hungarian Society for Human Genetics (26 founding members)	Budapest, Hungary
	5th annual EEMS meeting; panel discussion on "legal aspects for the safety evaluation of chemicals by mutagenicity tests"	Florence, Italy
	International School of General Genetics, 1st course: "Environmental Mutagenesis" (NATO Advanced Study Institute)	Florence, Italy
	International Symposium on New Developments in Mutagenicity Testing of Environmental Chemicals	Zinkovy Castle, Czechoslovakia
	International course on environmental mutagenesis	Erice, Italy
	Training course on mutagenicity testing	Bombay, India
	"Chemical mutagenesis: A survey of the 1973 literature" (EMIC report)	Oak Ridge, Tenn.
1976	Conference, "Birth Defects: Detection and Prevention"	Martinique
	"Genetic Toxicology Testing" begins as special section of *Mutation Research,* published quarterly	
	7th annual EMS meeting	Atlanta, Ga.
	Workshop, "Basic and Practical Approaches to Environmental Mutagenesis and Carcinogenesis"	Dushanbe, Tajikstan (USSR)
	Symposium, "Standarization and Development of Mammalian Test Methods"	Neuherberg, W. Germany

Year	Event	Place
	Workshop, "Design of New Methods for Assessing Mutations in Mammalian Systems" (proceedings published in *Genetics*)	
	East Asia workshop, "Mutagenicity Testing of Chemicals" (U.S.-Japan Cooperative Medical Science Program)	Mishima, Japan
	DHEW Subcommittee on Environmental Mutagenesis formed	
	Workshop, "In Vitro Metabolic Activation in Mutagenesis Testing"	Research Triangle Park, N.C.
	Symposium, "The Role of Metabolic Activation in Producing Mutagenic and Carcinogenic Environmental Chemicals," February 9–11	Research Triangle Park, N.C.
	"Chemical mutagenesis: A survey of the 1974/1975 literature" (EMIC report)	Oak Ridge, Tenn.
	Mutation Research: Problems, Results, and Perspectives published (Auerbach)	
	Indian Environmental Mutagen Society formed	
	Mutagenesis published (Drake)	
	Chemical Mutagens: Principles and Methods for Their Detection, vol. 4, published (Hollaender, ed.)	
	Congress passes Toxic Substances Control Act of 1976 (October 11)	

Sources: Most of the data for this table were collected through a systematic search of announcements and news items in key journals, newsletters, and annual reports. Monographs were located by searching the WorldCat electronic database (keywords: mutagens, mutagenesis). A smaller proportion of the data was collected at random, as I came across relevant information in the normal course of historical research. Although all source documents are in my possession, I have not listed them here individually.

Main Sources: Mutation Research, vols. 1–32 (1964–1976); *EMS Newsletter,* nos. 1–6 (1969–1972); *Environmental Mutagenesis and Related Subjects* (1973–1976); *Reviews in Genetic Toxicology* (1975–1976); *Genetic Toxicology Testing* (1976); National Institute of Environmental Health Sciences, *Annual Report,* "Environmental Mutagenesis Branch–Summary Statement"(1972–1976); ORNLBD, Annual and Semi-Annual Progress Reports (1965–1969); WorldCat electronic bibliographic database, http://newfirstsearch.oclc.org.

NOTES

CHAPTER 1   SITUATING GENETIC TOXICOLOGY

1. Minutes, Meeting of the Ad Hoc Committee of the Environmental Mutagen Society (January 8, 1969), EMS.

2. Original research on chemical mutagenesis increased 200–500 percent per year between 1968 and 1972 (Wassom 1973).

3. "Membership list," *EMS Newsletter*, 1969 (2):74–85, EMS.

4. Important federal-level developments included the National Institute of Environmental Health Science (NIEHS), established in 1969 to direct basic research on "the effects of environmental factors, singly and in the aggregate, upon the health of man" (Research Triangle Institute 1965), and expert panels convened at the behest of several federal government agencies and departments to address the biological impacts of synthetic environmental chemicals. These included a major report, "Pesticides and Their Relationship to Environmental Health," published by the Department of Health, Education, and Welfare in conjunction with the Department of Agriculture, and a report on food safety standards issued by the Food and Drug Administration (FDA) advisory committee panel on "Safety Evaluation of Food Additives and Pesticide Residues" (Food and Drug Administration Advisory Committee on Protocols for Safety Evaluation 1970; U.S. Congress 1969; U.S. Department of Health, Education, and Welfare 1969). For a history of U.S. environmental regulatory policy, see Andrews (1999).

5. The history of contemporary genomic-based environmental health sciences is just beginning to be written (Shostak 2003b). On industrial hygiene and occupational disease research and politics, see, for example, Corn (1992), Sellers (1997), Gottlieb (1993), Rosner and Markowitz (1987), and Markowitz and Rosner (2002). On environmental theories and politics of cancer, see Epstein (1979) and Proctor (1995).

6. Elizabeth S. Von Halle, former membership director, EMS. Personal correspondence (November 24, 1996).

7. Elizabeth S. Von Halle, 1997. EMS Membership Report, EMS, Oak Ridge, Tennessee. Author's files.

8. While much of this literature maintains a realist position toward the content of scientific knowledge, some work promotes a relativist or social constructionist epistemology through arguments about how scientific facts and theories are influenced by social processes (e.g., Knorr 1977).

9. Factors identified in this body of work range from the character of colleague relationships within research networks (Mullins 1976) and the structure of the research process (Knorr 1977) to broader institutional reforms in education and academic labor markets (Ben-David and Collins 1966) and the emergence of new consumer knowledge markets (Groenewegen 1987).

10. A powerful recent example of this style of analysis is Clark et al. (2003).

11. For empirical analyses of interdisciplines and interdisciplinary research, see Barmark and Wallen (1980); Bechtel (1986); Weingart and Stehr, eds. (2000); Gibbons et al. (1994); Lattuca (2001); and Abir-Am (1987, 1988, 1993). Most of the literature on "interdisciplinarity" derives from cultural and literary studies. In addition to Julie Klein's work and the edited volume by Messer-Davidow et al. cited earlier in the chapter, Moran's (2002) study is broadly representative of this much larger body of work.

12. *Ceteris paribus*, a cultural studies program, is more likely to hire faculty with Ph.D.s in geography or history or literature than from some other cultural studies program.

13. As in studies of trading zones (Galison 1996), standardized packages (Fujimura 1996), and boundary objects (Star and Griesemer 1989). Research on the "boundaries of science" owes much to Thomas Gieryn (1983, 1994, 1999). For studies of boundary making in explicitly interdisciplinary contexts, see Frickel (2004a) and Small (1999).

14. Minutes, EMS Council Meeting (September 18–19 and March 22, 1970), EMS.

15. On the social authority of scientists more broadly, see Rosenberg (1997) and Walters (1997).

16. Studies of biology, American or otherwise, that cover this period include Appel (2000), de Chadarevian and Kamminga (1998), Kay (2000), and Morange (2000).

17. On the role of biologists as Cold War public intellectuals, see Wolfe (2002).

18. Hollaender served as director of the Oak Ridge National Laboratory Biology Division from 1946 to 1966 and remained on in a consulting capacity until 1972. He made regular trips to Washington during this period and, according to my interviews, spent considerable time at the Cosmos Club, the city's elite and historic social club located near Dupont Circle. Daniel Greenberg (1999 [1967]:3) once described the Cosmos Club as the scientific community's cultural and intellectual equivalent to ancient Athens's Athenaeum.

19. As Kleinman and Vallas (2001) note, however, basic research in American universities was never the autonomous "ivory tower" policy makers, science boosters, and some historians and sociologists of science have maintained. On this and other "myths" of science policy, see Sarewitz (1996).

20. A few scientists I interviewed used exactly this phrase to describe the culture at Oak Ridge during the mid–1960s. Interview data.

21. Interview data.

22. Some of these same genetics students carried Mother Nature's coffin in their town's Earth Day parade. Interview data.

23. The National Environmental Policy Act (NEPA), passed by Congress on December 24, 1969, and signed into law by President Nixon on January 1, 1970, mandated the creation of the U.S. Environmental Protection Agency and the Council on Environmental Quality. Congressional environmental protection legislation subsequent to NEPA included the Clean Air Act (1970), the Resource Recovery Act (1970), the Water Pollution Control Act (1972), the Federal Insecticide, Fungicide, and Rodenticide Act Amendments (1972), and others.

24. More recently, for example, "risk society" theorists (Beck 1999) have expressed similar concerns.

25. Victor McCusick, a clinical geneticist appointed head of a new Division of Medical

Genetics at Johns Hopkins Medical School in 1957, and James V. Neel, a radiation and population geneticist who in 1956 founded and chaired the Department of Human Genetics at the University of Michigan, were important figures whose laboratories "loomed particularly large on the landscape" of human genetics in 1959 (Kevles 1985:ch. 15, quote from 233). In biochemistry, new chromatographic and electrophoretic methods facilitated research on biochemical variants among large populations, and in cytology, new techniques made for improved karyotype analysis (Kevles 1985:235, 249).

26. On the legitimating function of disciplinary origin accounts, see Abir Am (1985).

27. For more and less sympathetic critiques, respectively, see Lenoir (1997) and Mulkay (1979).

28. See also Clarke (1998:ch. 6). Other studies are less specific. Latour's (1988) analysis of "the Pasteurization of France" is set amid a backdrop of a hygienics movement that never quite comes into full view. Similarly, Fujimura (1996) counts cancer research advocacy groups among the many and varied institutional actors that contributed to the development of an "oncogene bandwagon" in cancer science, although we never see the movement, its organizations, or its activists.

29. Woodhouse and Breyman (2004) take a similar approach in their study of green chemistry. For related discussions of how medical and scientist activism is organized toward environmental justice, see McCally (2002) and Frickel (2004b).

30. My understanding of social movement dynamics owes much to the variously termed but basically similar "resource mobilization," "political process," and "contentious politics" perspectives. See McAdam et al. (1996), McAdam et al. (2001), and Tarrow (1998).

31. Kuznick (1987) describes scientists' mobilizations for social and political reform during the interwar period; however, most historiography on scientist political activism is set during the Cold War. For studies of scientist activism in reaction to the bomb and radiation fallout, see Smith (1965), Rotblat (1972), Kevles (1978), Boyer (1985), and Divine (1978). Jessica Wang (1999) examines scientists' retrenchment against anti-Communism. Jan Sapp (1987) and Audra Wolfe (2002) treat American biologists' opposition to Lysenkoism. Key sources for the relationship between American genetics and eugenics movements are Kevles (1985) and Paul (1995). On scientists' antiwar activism during the Vietnam War, see Moore (forthcoming). Interesting case studies of scientist activists in the policy arena and in the public sphere are found, respectively, in Primack and von Hippel (1974) and Goodell (1977).

CHAPTER 2   WORKING ON MUTATIONS

1. Other historians of genetics have raised similar sentiments, as in Ana Echeverria's (1995:89) more subtle observation that the challenge for geneticists historically has been "not so much to accept mutation as the generating process of variation, but rather to attempt to define its limits."

2. A recent text uses this analogy: "What a gene does . . . reveals its presence much as odors, sounds, and slight disturbances of its surroundings reveal the presence of a field mouse to an alert cat" (Wallace and Falkinham 1997:1).

3. Kohler (1994:20) argues that *Drosophila* first entered biologists' laboratories not as experimental material for genetics but as a pedagogical tool.

4. Each generation of progeny was also crossed with untreated flies.

5. As noted in the previous chapter, these projects still do not exhaust Muller's curriculum vitae for political activism, a political biography that earned him an enigmatic reputation among his colleagues then and historians now.

6. Carlson (1981:144–147) describes these experiments in greater detail.

7. Hessenbruch (2000) examines the economic history of x-ray science and technology.

8. A Jew, Auerbach immigrated to Edinburgh from Germany in 1935 and received her Ph.D. in 1937, studying developmental genetics under F.A.E. Crew. Crew assigned Auerbach to be a research associate to H. J. Muller during his time in Edinburgh in the late 1930s, and it was in conversation with Muller that Auerbach became interested in mutagenesis (Carlson 1981:246–247). Their correspondence in later life remained regular, if not close, and upon Muller's death Auerbach contributed a seven-page obituary "note" that was published in *Mutation Research* (Auerbach 1968).

9. Like radiation, mustard gas creates burns that tend to heal slowly, often breaking down again after healing. It was these characteristics that first suggested to the pharmacologist Robson that mustard gas might, like x rays, "inhibit cell division through direct action on the chromosome" (Auerbach et al. 1947:244).

10. During the war, the British government imposed a security ban on all information relating to "war gases" (Auerbach et al. 1947:244).

11. Beale (1993) notes that Auerbach and Robson experienced difficulty applying mustard gas to flies in nonlethal doses.

12. Lethal mutations prevent the development of individuals. "Sex-linked" lethals are lethal mutations on the X chromosome. "Crossing over" is a process whereby homologous chromosomes exchange corresponding parts. A "rearrangement" occurs when a chromosome segment is inverted (thus preventing crossing over).

13. Auerbach (1962a) reviews standard methods in *Drosophila*, the mouse, wasps, flowering plants, and various microorganisms.

14. Loveless (1966:v) describes alkylating agents as those compounds possessing the "capacity to alkylate biologically functional chemical groups *in vivo* under normal physiological conditions."

15. A few examples include the mouse specific-locus test for germ-cell mutagenesis (Russell 1994; Russell 1989), tests for forward and reverse mutations in the ad–3 region of *Neurospora crassa* (de Serres and Kolmark 1958), the *Tradescantia* stamen-hair system (Underbrink et al. 1973), and the *rII* locus test for bacteriophage T4 (Drake 1963). Many others could be noted.

16. At least one new method with great potential for finding mutagens was not incorporated into standard practice in chemical mutagenesis. The "chemostat" method, developed in 1950, warrants only brief mention in Auerbach's methodology textbook (1962b:109–110), nor is there a chapter devoted to this method in Hollaender and de Serres's ten-volume series, *Chemical Mutagens: Principles and Methods for Their Detection*. Nevertheless, the chemostat "offered not only an enormous increase in precision but a built-in test of many of the assumptions. Very small changes in the mutation rate could be measured, and therefore mild mutagens could be detected" (Crow 1989:9). In an interview, James Crow told me that he thought the reason the chemostat was not used for mutagenicity testing was economic: other sensitive systems were cheaper to run (James F. Crow, interview, Madison, Wisconsin, April 18, 1997). It is worth noting that the other methods Crow refers to were also easier for geneticists to make work.

17. As a woman working in a male-dominated profession, as an employee in a university agricultural research laboratory, as the author of a key discovery in genetics, and as a vocal proponent of a more holistic understanding of molecular biology than the one that has carried the day, Auerbach's life and research certainly invite close comparison to Cornell University geneticist Barbara McClintock (Comfort 2001; Keller 1983).

18. The term "chemosterilant" was coined by USDA entomologists in 1960 (LaBreque et al. 1960).

19. Even more appealing to the USDA entomologists and chemists than finding chemical means to sterilize laboratory flies was the possibility of identifying ecologically safe chemicals that could be administered directly to a natural insect population. Such compounds would eliminate the need for insect-rearing facilities altogether and significantly lessen the economic burden that accompanied the sterile-male release method (Knipling 1962).

20. In the United States, the Radiation Research Society was formed in 1952, its first annual meeting was held in 1953, and its journal began publication in 1954 (Failla 1954). The federal government's interest in radiation genetics was manifest mainly through research supported by the AEC and by a series of reports commissioned by the National Research Council (National Academy of Sciences–National Research Council 1956, 1960, 1972). International concern is illustrated by a series of United Nations reports published around the same time (e.g., United Nations 1958). On the political history of postwar radiation genetics in the United States, see Beatty (1988, 1993) and Lindee (1994).

21. The molecularization of the life sciences and its consonant reverberations throughout industry, politics, and culture have received considerable recent attention. See the book-length treatments on the rise of molecular biology by Kay (1993, 2000), Morange (2000), and Fujimura (1996), as well as the essays collected in Fortun and Mendelsohn (1999). On genetic engineering regulation and biotechnology, see Gottweis (1998), Wright (1994), and Thackray (1998).

22. Chemical mutagens did play an indirect role in some microorganism research. Chemical compounds such as acridine dyes were used to sensitize bacteria to ultraviolet radiation (Crow and Abrahamson 1965:282; see also Drake's discussion of acridine mutagens [1970:147]). As Auerbach (1976:8) notes, "It is true that chemical mutagens were applied to micro-organisms, but almost all important findings on chemical mutagenesis at that time came from experiments with *Drosophila*, while most important findings on UV-mutagenesis came from experiments on micro-organisms."

23. Mice were the main exception to this rule. Although research on mouse genetics didn't require larger populations, mice were, for obvious reasons, considerably more expensive than bacteria or yeast to breed, house, and study (Russell 1994).

24. Several scientists I interviewed expressed this opinion. Crow's remarks are representative:

    SF: If one had a laboratory set up for radiation genetics and one decided to study chemical mutagens, what kind of changes would have to take place?

    JC: Very little. You have to have an x-ray machine to do radiation. Chemicals are much simpler to study. You can take something off the shelf and the techniques to test it—growing bacteria or flies or whatever test system you're using—that didn't change. So it wasn't a big step in a research sense. (Crow, interview)

25. Despite the important influence of European scientists, the United States clearly dominated the field in terms of national output. U.S. scientists contributed nearly a

third (31 percent) of the total number of articles published in *Mutation Research* (1964–1968), and U.S. scientists made up 40 percent of the journal's forty-eight-person editorial board.

26. This is not to suggest that important mutation research was not going on elsewhere but only that the concentration of mutation research at these three institutions was at that time unequaled.

27. As Auerbach reminded one audience, "Once the mutant information has been firmly encoded in DNA, it has to be read, transcribed and put into effect, and a new biochemical pattern has to be evolved in the mutant cell" (1963:282).

28. I refer here primarily to the research and regulatory laboratories in the AEC, NIH, FDA, and, after 1970, EPA.

29. "Organization of Research," October 1965. ORNLBD.

CHAPTER 3   MAKING ROOM FOR ENVIRONMENTAL MUTAGENS

1. The term "mutational load" refers to the average accumulation or decline of deleterious germinal mutations in human populations. Because most germ-cell mutations are not extreme enough to kill or sterilize an individual, and because advances in hygiene and medical treatment have enabled genetically weakened individuals to pass those damaged genes along to offspring, Muller argued that mutations are not being regularly eliminated from the human gene pool as they would have been among our less technologically advanced ancestors. For the political and cultural context of Muller's arguments, see Paul (1987).

2. This interesting and prescient exchange is reprinted in Lederberg (1997). The quote is from page 6.

3. Lederberg shared more than a Nobel Prize (awarded in 1958 for work in bacterial genetics) with his older colleague Muller. A consummate disciplinary entrepreneur, Lederberg founded two university departments (the Department of Medical Genetics at the University of Wisconsin–Madison and the Department of Genetics at Stanford University) and worked to establish a new discipline for the scientific study of extraterrestrial life, or "exobiology" (Dick 1996). He was also a public intellectual who wrote a weekly column for the *Washington Post* from 1966 to 1971, "Science and Man," that explored various issues of concern to Lederberg involving science's relationship to society and social policy. Nearly 22 percent of his 231 columns addressed topics involving consumer protection and the environment, some specifically focused on radiation and chemical mutagens. A similar fraction addressed human genetics and reproduction, including several with eugenics-oriented themes (Wolfe 2002:80). Lederberg's columns can be found online at http://profiles.nlm.nih.gov/BB/Views/ AlphaChron/ alpha/10035/10000/.

4. My thanks to Audra Wolfe for suggesting the phrase "bully pulpit."

5. Neel and Schull were architects of the Atomic Bomb Casualty Commission study of survivor populations at Hiroshima and Nagasaki (Beatty 1988; Lindee 1994). Other notables among the twenty-three attendees included Curt Stern from UC-Berkeley and Milislav Demerec from Brookhaven National Laboratory. The conference proceedings are published in Schull (1962).

6. The three categories involved drugs and other substances to which (1) large populations are exposed to only occasionally, (2) smaller subpopulations are chronically exposed, and (3) large populations are chronically exposed. The first category could be assumed not to pose a genetic hazard; the second could be assumed to pose a

genetic hazard to only a small proportion of the population, and therefore "the genetic hazard to the race would be minimal." Substances in the third category represented the most immediate and significant concern. These included ethyl alcohol, nicotine, water additives (chlorine and fluoride), foods (e.g., mustard seeds) and food additives (e.g., nitrates and nitrites), insecticides, industrial air pollutants, oral contraceptives, contraceptive jellies and creams, theobromine, and caffeine (Goldstein 1962).

7. Muller's role in establishing genetic toxicology was indirect. He died in 1967 after a long period of poor health during which he was not physically capable of the kind of political and social activism for which he had by then become infamously known among colleagues and critics. Nevertheless, it seems likely that his personal connections and influence did play a role in piquing interest in chemical mutagenesis research at the National Research Council and at Oak Ridge (Lederberg 1997). The talk Muller gave on the public health implications of chemical mutagens at the FDA also seems to have had reverberating effects that Muller himself would not have predicted. Arriving at the FDA to head its Cell Biology Branch a year after Muller delivered his talk there, Marvin Legator recalled his "delight in discovering Dr. Muller's report [in 1963]," admitting in 1970 that "the presentation of Muller's, and my later correspondence with him greatly influenced the activities of the Cell Biology Branch at FDA" (1970:240). Those activities involved research on what Legator would in 1967 term "genetic toxicology."

8. Alexander Hollaender, interview by Ida C. Miller (February 2, 1982), RRS, MS 1709, Folder 2.

9. Ibid.

10. Alexander Hollaender, Acceptance Speech for the Distinguished Contribution Award, Society for Risk Analysis (1985), RRS, MS 1709, Folder 19; Hollaender to Harvey Pratt, School of Medicine, Laboratory of Radiobiology, University of California at San Francisco (May 14, 1976), RRS, MS 1261, Box 3, Folder 13; Hollaender to David Perkins, Department of Biological Science, Stanford University (January 3, 1978), RRS, MS 1709, Folder 18.

11. Interview data.

12. "He [Meselson] was quite an activist," Crow told me in our interview. "So much of what I wrote and what the group agreed with was based on what Meselson had said. He should get a lot of credit for this." Other interviewees confirmed this view of the important activist role that Meselson played in promoting chemical mutagenesis research in science policy and administration circles during the 1960s. Meselson's work in chemical and biological weapons (CBW) policy and his opposition to U.S. military chemical weapons use in Vietnam are summarized in Primack and von Hippel (1974:ch. 11). Meselson's comparatively modest involvement in the genetic toxicology movement is explained by the fact that between 1969 and 1971 he spent half his time on "anti-CBW activities" (148).

13. For his published account of this episode, see Crow (1989).

14. Crow described the journal to me as an "out of the way place" that arguably was not the most visible outlet for the report. "*Science* would have liked to have had it but by that time [1968]—I didn't realize that it would attract the kind of attention that it did—and I'd already committed to [*Scientist and Citizen*] and I thought that I should stay by that commitment." One wonders whether and how things might have happened differently had either NIH or a journal with the international prestige of *Science* published the report.

15. Interview data.

16. One scientist I interviewed advanced the theory that Roger Tory Peterson's *Field Guide to Birds of North America* (1934) was more integral to raising awareness of the ecological connections between wildlife habitat and pollution among scientists and naturalists than *Silent Spring.*

17. Examples include the Agricultural Research Service, Entomology Division; the Fish and Wildlife Service; the AEC; and the Public Health Service Water Quality Laboratories.

18. Nonnuclear work at Oak Ridge during the 1960s involved, for example, the development of a desalination plant, research on viruses and vaccines, the construction of a facility to study the relationship between aquatic and terrestrial ecosystems, and a leading role in the National Science Foundation's International Biological Program (Johnson and Schaffer 1994:123–124). On the latter, see Kwa (1987) and Bocking (1995).

19. Among the more significant mutagenicity bioassays originating in the Biology Division was a reverse-mutation system in Neurospora (associated with Frederick J. de Serres), the mouse specific-locus test (developed by William Russell), an in vitro Chinese hamster cell bioassay (developed by Ernest Chu), and the "S–9" metabolizing agent (developed by Heinrich Malling) incorporated into a number of in vitro bioassays, most notably the Salmonella/microsome bioassay commonly known as the Ames test.

20. Internal report, no date (probably 1976), 3, EMS.

21. The other information centers were the Accelerator Information Center, Actinide Research Information Center, Atomic and Molecular Processes Information Center, Biogeochemical Ecology Research Center, Charged Particle Cross Section Data Center, Civil Defense Research Collection, Engineering Data Collection, Information Center for Internal Exposure, Isotopes Information Center, Nuclear Desalination Information Center, Nuclear Safety Information Center, Office of Saline Water Materials Information Center, Photographic Research Collection, Radiation Shielding and Monitoring Information Center, and Research Materials Information Center. "Environmental Mutagen Information Center of the Environmental Mutagen Society" (June 4, 1970), John Wassom files, Oak Ridge, Tennessee.

22. Congress first allocated monies to fund a planning study for an "Environmental Health Sciences Center" in 1961.

23. In its earliest years, de Serres's branch was staffed mainly by scientists accompanying him from the Oak Ridge Biology Division's Fungal Genetics Section. The migration from Tennessee's Oak Ridge to North Carolina's newly christened Research Triangle was explained to me in terms of the "pull" of new opportunities to advance environmental mutagenesis at NIH in combination with the "push" of administrative politics at Oak Ridge following Alexander Hollaender's departure. Interview data.

24. For example, during 1975 EMB initiated contracts with geneticists at the Universities of Washington and Wisconsin, the Jackson Laboratory, and the Miles Laboratory, and it negotiated interagency agreements with the EPA (for base-line studies of the mutagenicity of airborne industrial chemicals) and with the Oak Ridge National Laboratory (to support data collection at EMIC) (NIEHS 1975b:151). In-house research was organized into four laboratory sections: microbial and plant genetics, biochemical genetics, somatic cell genetics, mammalian genetics, and population monitoring and epidemiology.

25. For example, through the U.S.-Japan Cooperative Medical Sciences Program, the U.S.-

USSR Environmental Protection and Environmental Health Agreements, and the U.S.-Germany Life Sciences Program and with the International Agency for Research on Carcinogenesis (IARC) in Lyon, France.

26. Mimeographed copy of William D. Ruckelshaus, "An Address to the American Society of Toxicology" (March 9, 1971), RRS, MS 1261, Box 3, Folder 15. A note affixed to the speech, from an EPA Office of Public Affairs staffer to FDA geneticist Marvin Legator reads, "I hear that Mr. Ruckelshaus' speech was not warmly accepted by the audience. I judge from that, that we did not miss the mark by far." Thomas F. Williams to Marvin Legator (March 10, 1971), RRS, MS 1261, Box 3, Folder 15.

27. Ribicoff to Hollaender (June 24, 1971); Hollaender to EMS Council (July 2, 1971); Minutes, EMS Council Meeting (July 27, 1971); Lindsay to Hollaender (June 22, 1971), EMS. Dale R. Lindsay was the associate commissioner for science and acting director of NCTR. That center's initial research focus was on large-scale animal studies of the effects of low-dose, long-term exposure to potentially toxic chemicals that were to be selected "according to man's exposure in foods and other aspects of his environment." NCTR Task Force, Working Group F, "Mutagenesis Protocols," RRS, MS 1261, Box 3, Folder 2. Both Flamm and Fishbein later took jobs at NCTR.

28. Minutes, Joint Meeting, EMS Executive and NAS ad hoc Subcommittee on Problems of Mutagenicity, Committee on Problems of Drug Safety (March 23, 1970); Minutes, EMS Council Meetings (March 24, 26, 1970); Minutes, EMIC Register Meeting (March 25, 1970); Minutes, EMS "Training" Meeting (March 28, 1970); Memo, Samuel S. Epstein to EMS Council (September 20, 1970); "Fundamentals of Mutagenicity Testing, Woods Hole, Massachusetts, July 12–18, 1970," EMS, Heinrich Malling Papers. The quoted passage is from an untitled statement by Alexander Hollaender (June 12, 1970), RRS, MS 1261, Box 3, Folder 10.

29. Minutes, EMS Council Meeting (March 24, 1970), EMS, Malling Papers; Alexander Hollaender (June 12, 1970), RRS, MS 1261, Box 3, Folder 10; Minutes, EMS Council Meeting (July 8, 1972), EMS, Malling Papers.

30. The Drug Research Board, for example, refused to sponsor a mutagenicity workshop at UC-Berkeley organized by Bruce Ames in 1972, arguing that the EMS should bear major responsibility for financing the workshop. Minutes, EMS Council Meeting (July 8, 1972), EMS, Malling Papers.

31. Samuel S. Epstein, telephone interview with author, July 14, 1999. Although Epstein's original draft of the bill was critically important in that it included the section on genetic effects and mutagenicity testing, he states that geneticists had little to do with pushing the act through Congress.

CHAPTER 4   A WAVE OF SCIENTIST COLLECTIVE ACTION

1. Epstein, interview.

2. News release (March 1, 1969), EMS.

3. This exponential growth was a general phenomena characterizing science during the 1950s and 1960s (Price 1963).

4. Classic illustrations of this approach are found in Hobsbawm and Rudé (1968) and Tilly (1978).

5. In cases of reoccurring phenomena such as the annual meetings of professional societies, I counted only the inaugural event. See Appendix B for details and sources.

6. The first genetic toxicology textbook did not appear until 1980 (Brusick 1980).

7. Interview data.

8. Sax was a retired professor of genetics at the University of Pittsburgh, where he conducted experimental research in chromosome cytology. In 1969 he was seventy-six years old.

9. Di Luzio had been an assistant secretary of the interior for water pollution control. In 1969 he worked in the private sector as the president of an electrical engineering firm in Las Vegas. "F. Di Luzio to S. Epstein" (April 10, 1969), RRS, MS 1167, Box 2, Folder 3.

10. The decision to make membership in EMS open to a concerned public was made in the first meeting of the Ad Hoc Committee. "Minutes, Meeting of the Ad Hoc Committee of the Environmental Mutagen Society" (January 8, 1969), EMS.

CHAPTER 5   FRAMING SCIENTIST ACTIVISM

1. At the time his article was published in 1968, Crow was serving on the editorial board of *Scientist and Citizen*, the journal of Barry Commoner's St. Louis–based organization Committee for Nuclear Information, and had previously contributed another article to the journal, on the ecological dangers of chemical pest control methods (Crow 1966). In 1971 Joshua Lederberg began service on the Natural Resources Defense Council original board of directors. And Sam Epstein (at Harvard Medical School) and Marvin Legator (at FDA) both were involved in consumer safety research and policy reform efforts.

2. Frederick J. de Serres, mimeographed lecture (no date) 4, RRS, MS 1261, Box 3, Folder 15.

3. Alexander Hollaender. "Thoughts on pollution" (March 11, 1970), RRS, MS 1261, Box 3, Folder 16.

4. "Not too many years ago," Joshua Lederberg (1969) wrote in the same "Science and Man" column cited above, "I was able to compartmentalize my own thinking to such a degree that I did not immediately grasp the relationship between an abstraction, like the statistics of 'lethal mutations' in fruit flies, and the human impact of malformation in the new-born."

5. "1. Only germ cell mutations affect subsequent generations. 2. Most mutations are harmful. 3. Any increase in the mutation rate is by its nature difficult to detect. 4. Mild genetic effects are more numerous than severe effects. 5. Recessive mutations may remain in a population for hundreds of generations" (Crow 1971a:22–23).

6. Alexander Hollaender, "Opening Remarks, Symposium on Environmental Pollutants." Annual Meetings of the Radiation Research Society, Dallas, Texas (March 2, 1970), RRS, MS 1261, Box 3, Folder 16.

7. de Serres, mimeographed lecture.

8. Alexander Hollaender, no title (1969), RRS, MS 1261, Box 3, Folder 15.

9. Alexander Hollaender, "Genetic implications of pollutants" (April 1970), RRS, MS 1261, Box 3, Folder 2. The four-page paper was included in an Office of Science and Technology report on the need for a World Health Organization environmental health program. See "Known and suspected effects of environmental exposures on human health and well-being" (July 13, 1970), RRS, MS 1261, Box 3, Folder 15. This was a common theme in Hollaender's writing during this period. See also his "Aspects of environmental health planning" (January 16, 1970), RRS, MS 1261, Box 3, Folder 15, and "Draft thoughts on environmental studies" (January 16, 1970), RRS, MS 1261, Box 3, Folder 16.

10. Nearly every document I examined mentioned this gap between research and regulatory policy.

11. Samuel S. Epstein, "The role of the university in relation to consumer, occupational and environmental problems," 5. Transcript of talk given at Case Western University (January 15, 1971), RRS, MS 1261, Box 3, Folder 15.

12. Epstein, "The role of the university," 8.

13. Ibid., 2.

14. This claim is based on my review of course catalogs at the University of Wisconsin–Madison, Cornell University, University of California–Riverside, and University of California–Davis. Toxicology courses at these universities during the 1960s did not list genetics as a prerequisite for enrollment.

15. Alexander Hollaender, "General discussion" (January 28, 1974), RRS, MS 1261, Box 1, Folder 4.

16. This critique was a common one during the 1960s (Winner 1977). Carson's anticorporate version of environmentalist critique was not, however, generally shared by genetic toxicology scientist-activists. As Hollaender and others recognized early on, the participation of industry scientists was crucial to their cause, and alienating them with inflammatory rhetoric would serve few if any useful short-term goals. See Hollaender, "General discussion."

17. de Serres, mimeographed lecture.

18. Legator, untitled manuscript.

19. de Serres, mimeographed lecture. See also the exchange published in the letters section of the *EMS Newsletter* (Legator and Epstein 1970; Zbinden 1970).

20. Carson was not protesting the destruction of agricultural pests per se but rather the ecological imbalances brought on by industrial monoculture crop production. "All this is not to say there is no insect problem and no need of control. I am saying, rather, that control must be geared to realities, not to mythical situations, and that the methods employed must be such that they do not destroy us along with the insects" (Carson 1962:19).

21. Others were honored guests. Charlotte Auerbach suffered serious burns and rashes from the mustard compounds she used in her mutagenesis experiments in the 1940s (Beale 1993).

22. The 1960 Macy Conference on Genetics (Schull 1962), the 1966 NIH Genetics Study Section symposium (Crow 1968), and the 1969 Public Health Services report, *Pesticides and Their Relationship to Environmental Health* (U.S. Department of Health 1969), were the most frequently mentioned sources for these reviews.

23. For specific genetic disorders, mutations occur once in roughly 10,000 to 100,000 births. A controlled study that generated statistically significant results might have involved upward of twenty million people (Sanders 1969a:57).

24. Epstein, "The role of the university," 9–10; Hollaender, "Aspects of environmental health planning."

25. "Euphenics" is a term coined by Lederberg to refer to the application of molecular biology in progress toward "man's control of his own development," through manipulation during infancy or in utero of body organs such as the brain or, more practically, through advances in organ transplantation technologies (Lederberg 1963b).

26. Muller, Lederberg, and Crow (among many others) argued that the development of

suitable gene manipulation technologies, while foreseeable, remained a long way off. Muller urged the extensive use of available technologies and practices, namely the establishment of sperm banks, artificial insemination programs, and public education ("genetic counseling") as the most immediately feasible way to prevent genetic disease (1965). Lederberg argued that organ transplant technologies provided an alternative route to diminishing the effect of genetic disease (1963a). Crow supported both of these strategies (1965).

27. "Positive" eugenics refers to the selection of desired characteristics or traits (through, for example, artificial insemination). "Negative" eugenics refers to the elimination of undesired characteristics or traits (through, for example, forced sterilization) (Paul 1984:568, n. 3).

28. The first quote is from Gary Flamm's testimony at the Senate hearings on "Chemicals and the Future of Man" (U.S. Senate 1971:27); the second quote is from de Serres, mimeographed lecture.

CHAPTER 6   ORGANIZING A SCIENTISTS' MOVEMENT

1. This is the phrase Hollaender often used in writing about the origins of the EMS. As far as I have been able to ascertain, however, no one in this small group had graduate training in toxicology per se. It is doubtful whether the bench research that any of them regularly engaged in was toxicological in the common sense of measuring acute toxicity in higher-order animal systems. On the other hand, the research of almost all of these scientists could reasonably be called genetics.

2. In an interview, Hollaender named Epstein, Legator, and Nichols as "the only ones who backed me" in his early efforts to promote environmental mutagenesis. Hollaender, interview. Hollaender, Meselson, Legator, de Serres, Freese, Malling, and Epstein were present at the founding meeting held in New York City in early January 1969. Minutes, "Ad hoc committee of the Environmental Mutagen Society (January 8, 1969), EMS, Malling Papers. The Institute for Medical Research was a private biomedical laboratory located in Camden, New Jersey.

3. This sentiment seems to be ubiquitous among those who knew and worked with Hollaender. Descriptions of his personality traits are recited time and again in articles on the history of genetic toxicology (Wassom 1989), in published tributes to Hollaender (e.g., Setlow 1968; von Borstel and Steinberg 1996), and in numerous interviews that I conducted. The characterization offered on the occasion of Hollaender's reception of the Enrico Fermi Award by Alvin Weinberg, who was director of the Oak Ridge National Laboratory during Hollaender's tenure as Biology Division director, is not atypical: "He could cajole, he could threaten, he could implore, he could serve as role-model—he was both a kindly and a stern father figure—fiercely protective of his flock in dealing with rather superfluous administrators (like the laboratory director), fiercely demanding of the people around him." Alvin Weinberg, "A tribute to Alexander Hollaender" (May 3, 1984), RRS, MS 1709, Folder 19.

4. RRS, MS 1261, Box 3, Folders 21–30. These files contain detailed records of Hollaender's foreign travel in his capacity as director of the Biology Division from 1961 to 1972.

5. Between 1961 and 1983, Hollaender organized sixteen international symposiums on a range of topics in modern biology that convened in various Latin American countries. Richard Setlow, unpublished speech nominating Alexander Hollaender for the Enrico Fermi Award (May 11, 1983), RRS, MS 1709, Folder 18.

6. Alvin Weinberg later noted that Hollaender was "the first administrator of the new style of biological research—Big Biology. Hollaender alone had the vision to recognize that a great biology laboratory could be set up in the hills of East Tennessee . . . and would become the world's foremost center for radiation biology—and one of the world's great laboratories for basic and applied biological research." Weinberg, "A tribute."

7. Daniel Billen, "A history of the University of Tennessee—Oak Ridge Graduate School of Biomedical Sciences," unpublished manuscript (February 1990) 1, ORNLBD.

8. Hollaender to Philip Handler (July 27, 1965), RRS, MS 1261, Box 1, Folder 24.

9. These projects are briefly enumerated in a letter from Hollaender to Philip Slater, then president of the Anderson Foundation. Hollaender to Slater (April 13, 1970), RRS, MS 1709, Folder 18.

10. Hollaender to D. B. Dill (January 5, 1966), RRS, MS 1261, Box 1, Folder 23.

11. Hollaender's career in science administration continued throughout his working life. In 1973 he left Oak Ridge, moving to Washington, D.C., to work as a research consultant for Brookhaven National Laboratory under the auspices of Associated Universities, a consortium of nine prominent research universities in the northeastern United States. Hollaender to Auerbach (January 29, 1975), RRS, MS 1261, Box 1, Folder 20. In 1973 Hollaender founded the Council for Research Planning in the Biological Sciences, serving as that organization's president until his death in 1986.

12. The questionnaire was distributed to ninety-five scientists. Sixty-three responded. Of those, twenty-nine favored "a formal and independent group," fourteen favored "a formal and independent Society," and thirteen favored "a formal Society sponsored by an agency." Interestingly, there was no option for *not* forming some type of organization. Questionnaire (January 21, 1969), EMS, Malling Papers.

13. At the time of transfer, the stock was valued at $10,000. Hollaender to Joseph Slater (May 28, 1969), RRS, MS 1261, Box 3, Folder 3; Minutes, EMS Council Meeting (September 18–19, 1970), EMS, Malling Papers; Hollaender, interview.

14. To date, most studies of boundary organizations have examined national or international governmental organizations, such as those involved in technology transfer (Guston 1999, 2000) or global climate change (Miller 2001), or quasi-independent organizations involved directly in science policy, such as the International Research Institute for Climate Prediction at Columbia University (Agrawala et al. 2001). Kelly Moore's (1996) study of "public interest science organizations" and Abbey Kinchy's and David Kleinman's (Kinchy and Kleinman 2003; forthcoming) research on the Ecological Society of America are two key exceptions.

15. Minutes, EMS Council Meeting (March 22, 1970); Minutes, EMS Business Meeting (March 23, 1970); and Minutes, EMS Council Meeting (September 18–19, 1970), EMS, Malling Papers. Six years later expansion into other closely related areas of research interest remained a contested issue. "I strongly believe that, while we should keep an eye on developments [in environmental teratogenicity], we should not contemplate making a major investment of time and energy here. . . . I am impressed by the rather poor correlation between mutagenicity and teratogenicity, and see no need for us to get in the birth defects arena per se." Memo, John Drake to Frederick J. de Serres (February 6, 1976), EMS, Malling Papers. John Drake was president of the EMS at the time he wrote this memo to de Serres, who served as chair of an EMS "Long-Range Planning Committee."

16. Alexander Hollaender, "Thoughts on environmental studies," RRS, MS 1261, Box 3, Folder 16.

17. Interview data. Several scientists I interviewed made essentially the same point.

18. News release (March 1, 1969), EMS, Malling Papers (emphasis added).

19. Minutes, EMS (February 8, 1969), EMS, Folder "EMIC Archives."

20. Hollaender to Epstein (March 4, 1969), RRS, MS 1261, Box 3, Folder 9. This letter contains a preliminary list of fifteen journals in which Hollaender thought EMS announcements should be published. It does not reflect all of the journals contacted but does indicate the range of audiences Hollaender hoped to attract.

21. Minutes, EMS Council Meeting (March 22, 1970), EMS, Malling Papers.

22. Malling to Walne (September 24, 1970), EMS, Malling Papers. This sort of local publicity (Malling worked in the Biology Division at Oak Ridge) was not insignificant. At least two people I interviewed had come to genetic toxicology and the EMS directly through the University of Tennessee.

23. Alexander Hollaender (December 6, 1970), RRS, MS 1261, Box 3, Folder 10.

24. Program, Second Annual Meeting of the Environmental Mutagen Society, RRS, MS 1261, Box 3, Folder 6. See also Minutes, EMS Council Meeting (September 18, 1970), EMS, Malling Papers.

25. Program, Fifth Annual Meeting of the Environmental Mutagen Society (March 8–11, 1974), RRS, MS 1261, Box 3, Folder 6.

26. Radiation Research Society Meeting Program (March 2–5, 1970), MS 1167, Box 2, Folder 3. See also Hollaender to Laughlin (November 14, 1969), Hollaender to Haagen-Smit (November 21, 1969), Hollaender to Epstein (November 11, 1969), RRS, MS 1105, Box 3, Folder "Radiation Research Society March 2–5, 1970, Dallas."

27. Minutes, EMS Council Meeting (July 27, 1971), EMS, Malling Papers. See also Chu to de Nova, Chu to Hollaender, Chu to de Serres (all October 24, 1975), RRS, MS 1261, Box 1, Folder 18.

28. Hoffman-LaRoche, Abbott Labs, and Squibb each donated $500 toward the expenses incurred at the first EMS meeting. Minutes, EMS Council Meeting (March 22, 1970). Legator to Hollaender (October 22, 1971), typed list containing thirty-nine chemical companies and handwritten note from Hollaender's secretary that reads: "This is apparently a list of chemical companies from Legator." RRS, MS 1261, Box 3, Folder 7. Minutes, EMS Council Meeting (March 26, 1972), EMS, Malling Papers.

29. Hollaender, interview.

30. Kinchy and Kleinman (forthcoming) find a similar dynamic in a historical analysis of the Ecological Society of America's environmental policy positions (see also Nelkin 1977). Their study does not, however, examine internal documents that may have revealed considerable conflict and negotiation within the organization. Consequently, their conclusion that professional scientific societies are unlikely sources for progressive political change in science, while intuitive, may be premature. For a suggestive counterexample, see Woodhouse and Breyman (forthcoming).

31. EMS report to IRS (draft, no date), RRS, MS 1261, Box 3, Folder 9 "EMS Legal Correspondence (1969)."

32. As an EMS lawyer explained to Samuel Epstein, tax law at the time precluded organizations with tax-exempt status from "utilizing a 'substantial part' of its activities in 'attempting to influence legislation,'" but that "an organization will not fail to qualify

[for tax exemption] 'merely because it advocates, as an insubstantial part of its activities, the adoption or rejection of legislation.'" The lawyer cautioned Epstein to "be wary of any participation in a public campaign during the adoption or rejection of specific legislation." Blinkoff to Epstein (October 2, 1969), RRS, MS 1261, Box 3, Folder 9 "EMS Legal Correspondence (1969)."

33. Minutes, EMS Council Meeting (July 27, 1971), EMS, Malling Papers.

34. Hollaender to Sobels (April 29, 1976), RRS, MS 1261, Box 3, Folder 13.

35. Individual EMS members could and did. Lederberg, for example, was on the board of directors of the National Resources Defense Council. Because he was not formally among the EMS leadership, his opinion about whether he favored an action plan was not recorded in the EMS Council Meeting minutes.

36. Minutes, EMS Council Meeting (October 17, 1972), EMS, Malling Papers.

37. Zeiger to Drake (November 5, 1976), EMS.

38. The survey administered by the ad hoc committee (see note 12 in this chapter) indicated that respondents considered the creation of a chemical registry of highest priority—considerably more important than the formation of a new society. EMIC was governed via two overlapping advisory committees that set goals and approved EMIC's annual budget. One was composed of EMS members from different disciplinary backgrounds, and the other was composed of representatives of the various government agencies that provided funding for EMIC. "Answers to SEQUIP questionnaire" (March 27, 1970), John Wassom, personal files.

39. EMIC Annual Report to EMS Council (March 22, 1971), John Wassom, personal files.

40. Minutes, EMS Council Meeting (March 26, 1972), EMS, Malling Papers.

41. "Answers to SEQUIP questionnaire" (March 27, 1970), John Wassom, personal files.

42. EMIC Annual Report to EMS Council (March 22, 1971), John Wassom, personal files.

43. In one letter, written in June 1970, Malling mentions that he has been answering questions on mutagenicity at "a rate of one per day," even though EMIC was not slated to begin official operations until the following September. Malling to Peters (June 4, 1970), EMS, Malling Papers.

44. Minutes, EMIC Register Meeting (March 25, 1970); Minutes, EMS Council Meeting (March 26, 1972), EMS, Malling Papers; and Minutes, EMIC Program Committee (December 18, 1970), John Wassom, personal files.

45. Disinterestedness has been a common theme for geneticists, for example, in their movements to oppose Lysenko, radiation fallout, and even racism.

46. Malling to Sobels (October 1, 1970), EMS. Ethyl methanesulfonate and methyl methanesulfonate were standard alkylating agents used in routine chemical mutagenesis research. Cyclophosphamide, a known teratogen, was a compound used in cancer chemotherapy.

47. For fiscal year 1971, for example, EMIC was funded with a total budget of $40,000, with the NIEHS, FDA, NSF, and AEC each contributing $10,000. Minutes, EMS Council Meeting (September 18–19, 1970), EMS, Malling Papers.

48. Minutes, EMS Council Meeting (July 8, 1972), EMS, Malling Papers.

49. Minutes, EMS Council Meeting (March 26, 1972), EMS, Malling Papers. A budget shortage in 1971 forced EMIC to curtail many of its data-collection efforts and sent EMS officers scrambling to locate additional "emergency" funds to keep the center running through the year. See Hollaender to Ruckelshaus (January 4, 1971); Kissman to Davis (January 11, 1971); Memo, Malling to EMIC Staff (February 11, 1971), all in John Wassom, personal files.

50. My interpretation here runs counter to Latour's general theory of interest translation (Latour 1987:108–121; 1988:65–67). In this case, interested groups like the Environmental Defense Fund and the Association of Analytical Chemists were purposefully *not* translated. Here we see how the *absence* of an association in this particular context became the mechanism that funneled power into the EMS "actor-network."

51. Minutes, EMS (February 8, 1969), EMS, "EMIC Archives."

52. Ibid. Cyclamate was an artificial sweetener banned by the FDA in 1969; hycanthone was a drug used to combat the tropical disease schistosomias; nitrosamines were found in food preservatives.

53. Minutes, EMS Council Meeting (September 18–19, 1970), EMS, Malling Papers; Nichols to Sparrow (October 2, 1971), RRS, MS 1261, Box 4, Folder 12; Nichols to Mooreland (October 2, 1971), RRS, MS 982, Folder "Environmental Mutagen Society."

54. Minutes, EMS Council Meeting (October 17, 1972), EMS, Malling Papers. The Delaney Clause, passed in 1958 as part of an amendment to the Food and Drug Act, banned food additives that tested positive in human or animal carcinogenicity tests. Chu to de Serres (October 24, 1975), RRS, MS 1261, Box 1, Folder 18.

55. Not surprisingly, there were committees created for this as well. In 1972 EMS formed a subcommittee to plan "for training workshops on a regular, ongoing basis" that were to be "directed at the bench scientist level" and which would "concern themselves with problems of interpretation besides techniques." Minutes, EMS Council Meeting (March 28, 1972), EMS, Malling Papers.

56. Marvin Legator, "Workshop at Brown University for testing and evaluating chemicals for mutagenicity" (no date) 3. RRS, MS 1261, Box 3, Folder 2.

57. Representative data on genetics departments and graduate-level genetics curricula are difficult to track down. Graduate-level textbooks published during the 1960s and early 1970s do, however, support the contention that while mutagenesis as a set of basic phenomena was an important part of a genetics course work, the environmental and/ or public health implications of chronic exposure to chemical mutagens were not.

58. In 1977 the NIEHS instituted extramural training programs in environmental toxicology, environmental pathology, environmental epidemiology and biostatistics, and environmental mutagenesis. The latter program, the smallest of the group, involved "minor parts" of several institutional awards in environmental toxicology at universities and medical schools receiving NIEHS training grants and two postdoctoral fellowships (NIEHS 1977:13).

59. Program, "Workshop on Mutagenicity" (July 26–28, 1971), Brown University; Program, "International Workshop on Mutagenicity Testing of Drugs and Other Chemicals" (October 2–5, 1972), University of Zurich. Both in RRS, MS 1261, Box 3, Folder 2.

60. Legator, "Workshop at Brown University."

61. Hollaender to Garin (May 11, 1976), RRS, MS 1261, Box 3, Folder 13.

62. Ibid. See also Hollaender to Legator (March 23, 1976), Hollaender to Brown (February 23, 1976), Hollaender to Tazima (April 28, 1976), all in RRS, MS 1261, Box 3, Folder 13.

63. This situation provides a wonderful opportunity for a comparative analysis of knowledge production systems, which, unfortunately, is beyond the scope of the present study.

64. Interview data.

CHAPTER 7   CONCLUSION

1. The assay is the 7-locus specific locus assay in mice developed by William Russell at Oak Ridge. David DeMarini, personal communication, August 9, 2003.

2. EMS-related societies in the United States are the Genotoxicity and Environmental Mutagen Society, the Genetic Toxicology Association, and the Genetic and Environmental Toxicology Association (http://www.ems-us.org/, accessed August 4, 2003).

3. www.iaems.org.nz/members, accessed August 5, 2003.

4. Interview data.

5. For example, a rudimentary search for graduate programs featuring genetic toxicology conducted in the online Peterson's Guide to Graduate Programs (keyword phrase: "genetic toxicology") resulted in 202 hits (http://www.petersons.com/, accessed August 4, 2003). The online database Proquest Digital Dissertations contains listings for eighty-five dissertations containing the phrase "chemical mutagenesis" in the title or abstract and sixteen containing the phrase "genetic toxicology" (http://wwwlib.umi.com/dissertations/gateway, accessed August 4, 2003).

6. DeMarini, personal communication.

7. As Star and Griesemer (1989:393) describe them, "boundary objects are objects which are both plastic enough to adapt to local needs and constraints of the several parties employing them, yet robust enough to maintain a common identity across sites." In genetic toxicology, mutagenicity bioassays became a means of translation among geneticists, pharmacologists and toxicologists, and regulatory agencies.

8. A Genetic Toxicology Testing Program began at NIEHS in 1978. The EPA-sponsored Gene-Tox program, which ran for much of the 1980s, can be credited as the first major effort to systematically evaluate and organize the mutagenicity data that EMIC collected (Waters 1979).

9. Later research showed that the predictive correspondence was closer to 70 percent than the near-perfect correlation Ames initially estimated (Tennant et al. 1987).

10. The authors list as examples of "confounding factors" age, diet, sex, smoking status, illness, recent medication, and radiation exposure.

11. The role of academic professors in environmental radicalism is not well understood, but some did not shy away from contentious politics. Norman Sanders, an assistant professor of geography at the University of California—Santa Barbara, taught courses on "environmental defense" that clearly went beyond the reach of environmental policy debate. The book Sanders (1972) wrote in conjunction with the course is titled *Stop It! A Guide to Defense of the Environment* and includes chapters titled "Forming a Group," "Getting the Message Out," and "Offensive and Defensive Environmental Law (including what to do when arrested)."

12. In many respects, Epstein is a consummate activist in science, with a curriculum vitae that lists press releases, membership in public interest organizations, media interviews, and congressional and agency testimony, in addition to more than 250 scientific publications (http://www.preventcancer.com/curriculum.pdf, accessed August 8, 2003).

13. Denny Larson, coordinator of the National Refinery Reform Campaign and Anne Rolfes, director of the Louisiana Bucket Brigade, personal communication, June 7, 2003.

14. http://www.fas.harvard.edu/%7Ehsp/.

15. James F. Crow, curriculum vitae. Author's files.

16. NexisLexis Congressional database; http://web.lexis-nexiscom/congcomp/.

17. Herman Brockman, "Role of genetic toxicology in environmental toxicology." Invited lecture, University of Wisconsin–Madison (February 2, 1983), Herman Brockman, personal files.

18. DeMarini, personal communication.

BIBLIOGRAPHY

ABBREVIATIONS

EMS   Archives of the Environmental Mutagen Society. Human Genome and Toxicology Group, Life Sciences Division, Oak Ridge National Laboratory, Oak Ridge, Tennessee.

ORNLBD   Oak Ridge National Laboratory Biology Division Library, Oak Ridge, Tennessee.

RRS   Radiation Research Society Archives, Hoskins Library, Special Collections, University of Tennessee, Knoxville.

Abir-Am, Pnina. 1985. "Themes, genres and orders of legitimation in the consolidation of new scientific disciplines: Deconstructing the historiography of molecular biology." *History of Science* 23:73–117.

———. 1987. "The biotheoretical gathering, trans-disciplinary authority and the incipient legitimation of molecular biology in the 1930s: New perspective on the historical sociology of science." *History of Science* 25:1–69.

———. 1988. "The assessment of interdisciplinary research in the 1930s: The Rockefeller Foundation and physico-chemical morphology." *Minerva* 26:153–176.

———. 1993. "From multidisciplinary collaboration to transnational objectivity: International space as constitutive of molecular biology, 1930–1970." In *Denationalizing Science: The Contexts of International Scientific Practice*, edited by Elisabeth Crawford, Terry Shinn, and Sorlin Sverker, 153–186. Boston: Kluwer Academic Publishers.

Agrawala, Shardul, Kenneth Broad, and David H. Guston. 2001. "Integrating climate forecasts and societal decision making: Challenges to an emergent boundary organization." *Science, Technology, and Human Values* 26:454–477.

Alexander, Peter. 1960a. "Mutation-producing chemicals." New Scientist 8:1073–1704.

———. 1960b. "Radiation-imitating chemicals." *Scientific American* 202:99–108.

Allen, Garland E. 1975. "The introduction of *Drosophila* into the study of heredity and evolution, 1900–1910." *Isis* 66:322–333.

Ames, Bruce. 1971. "The detection of chemical mutagens with enteric bacteria." In *Chemical Mutagens: Principles and Methods for their Detection*, vol. 1, edited by Alexander Hollaender, 267–279. New York: Plenum Press.

Ames, Bruce N., William E. Durston, Edith Yamasaki, and Frank D. Lee. 1973. "Carcinogens are mutagens: A simple test system combining liver homogenates for activation and bacteria for detection." *Proceedings of the National Academy of Science* 70: 2281–2285.

Amsterdamska, Olga. 1987. "Intellectuals in social movements: The experts of solidarity." In *The Social Direction of the Public Sciences*, edited by S. Blume, J. Bunders, L. Leydesdorff, and R. Whitley, 153–186. Dordrecht, Boston, and London: D. Reidel.

Andrews, Richard N. L. 1999. *Managing the Environment, Managing Ourselves: A History of American Environmental Policy*. New Haven, Conn.: Yale University Press.

Appel, Toby A. 2000. *Shaping Biology: The National Science Foundation and American Biological Research, 1945–1975*. Baltimore: Johns Hopkins University Press.

Armendares, S., and R. Lisker, eds. 1977. *Human Genetics: Proceedings of the Fifth International Congress of Human Genetics* (October 10–15, 1976; Mexico City). Amsterdam: Excerpta Medica.

Atwood, Kim. 1962. "Problems of measurement of mutation rates." In *Mutations, Second Macy Conference on Genetics*, edited by William J. Schull, 1–77. Ann Arbor: University of Michigan Press.

Auerbach, Charlotte. 1949. "Chemical induction of mutations." *Proceedings of the 8th International Congress of Genetics* (supplement volume):128–147.

———. 1961. *The Science of Genetics*. New York: Harper and Brothers.

———. 1962a. "Mutagenesis, with particular reference to chemical factors." In *Mutations, Second Macy Conference on Genetics*, edited by William Schull, 78–166. Ann Arbor: University of Michigan Press.

———. 1962b. *Mutations: An Introduction to Research on Mutagenesis*. Edinburgh and London: Oliver and Boyd.

———. 1963. "Past achievements and future tasks of research in chemical mutagenesis." In *Genetics Today (Proceedings of the XI International Congress of Genetics, the Hague, the Netherlands)*, edited by S. J. Geerts, 78–166. New York: Pergamon Press.

———. 1967. "Changes in the concept of mutation and the aims of mutation research." In *Heritage from Mendel (Proceedings of the Mendel Centennial Symposium Sponsored by the Genetics Society of America)*, edited by R. Alexander Brink, 67–80. Madison: University of Wisconsin Press.

———. 1968. "Obituary note: H. J. Muller 1890–1967." *Mutation Research* 5:201–207.

———. 1976. *Mutation Research: Problems, Results, and Perspectives*. London: Chapman and Hall.

———. 1978. "Forty years of mutation research: A pilgrim's progress." *Heredity* 40:177–187.

Auerbach, Charlotte, and John M. Robson. 1944. "Production of mutations by allyl *iso*thiocyanate." *Nature* 154:81.

———. 1946. "Chemical production of mutations." *Nature* 157:302.

———. 1947. "The production of mutations by chemical substances." *Proceedings of the Royal Society (Edinburgh Section B)* 62:271–283.

Auerbach, Charlotte, John M. Robson, and J. G. Carr. 1947. "The chemical production of mutations." *Science* 105:243–247.

Barmark, Jan, and Goran Wallen. 1980. "The development of an interdisciplinary project." In *The Social Process of Scientific Investigation*, edited by Karin D. Knorr, Roger Krohn, and Richard Whitley, 221–235. Dordrecht, Boston, and London: D. Reidel.

Beale, G. 1993. "The discovery of mustard gas mutagenesis by Auerbach and Robson in 1941." *Genetics* 134:393–399.

Beatty, John. 1988. "Genetics in the Atomic Age: The Atomic Bomb Casualty Commission, 1947–1956." In *The American Development of Biology*, edited by Ronald Rainger, Keith R. Benson, and Jane Maienschein, 284–323. Philadelphia: University of Philadelphia Press.

———. 1993. "Scientific collaboration, internationalism, and diplomacy: The case of the Atomic Bomb Casualty Commission." *Journal of the History of Biology* 26:205–231.

Bechtel, William, ed. 1986. *Integrating Scientific Disciplines*. Dordrecht: Martinus Nijhoff.

Beck, Ulrich. 1999. *World Risk Society*. Malden, Mass.: Blackwell.

Ben-David, Joseph. 1971. *The Scientist's Role in Society: A Comparative Study.* Chicago: University of Chicago Press.

———. 1991. *Scientific Growth: Essays on the Social Organization and Ethos of Science.* Berkeley and Los Angeles: University of California Press.

Ben-David, Joseph, and Randall Collins. 1966. "Social factors in the origins of a new science: The case of psychology." *American Sociological Review* 31:451–465.

Benford, Robert D., and David A. Snow. 2000. "Framing processes and social movements: An overview and assessment." *Annual Review of Sociology* 26:611–639.

Benson, Keith R., Jan Maienschein, and Ronald Rainger, eds. 1991. *The Expansion of American Biology.* New Brunswick, N.J.: Rutgers University Press.

Bocking, Stephen. 1995. "Ecosystems, ecologists, and the atom: Environmental research at Oak Ridge National Laboratory." *Journal of the History of Biology* 28:1–47.

Borkovec, Alexej B. 1962. "Sexual sterilization of insects by chemicals." *Science* 137:1034–1037.

Bourdieu, Pierre. 1975. "The specificity of the scientific field and the social conditions of the progress of reason." *Social Science Information* 14:19–47.

Boyer, Paul. 1985. *By the Bomb's Early Light: American Thought and Culture at the Dawn of the Atomic Age.* New York: Pantheon.

Brickley, Peg. 2002. "Matthew Meselson: Scientist and world statesman." *Scientist* 16 (www.the-scientist.com/yr2002/mar/features_020318.html).

Bridges, Bryn A. 1971. "Environmental genetic hazards: The impossible problem." *EMS Newsletter* 5:13–15.

———. 1993. "Obituary: Frederik Hendrik Sobels, 1922–1993." *International Journal of Radiation Biology* 64:469–470.

Brown, Kathryn S. 2000. "A new breed of scientist-advocate emerges." *Science* 287:1192–1195.

Brown, Phil, Stephen Zavestoski, Sabrina McCormick, and Brian Mayer. Forthcoming. "Health social movements: Uncharted territory in social movement research." *Sociology of Health and Illness.*

Brulle, Robert J. 1996. "Environmental discourse and social movement organizations: A historical and rhetorical perspective on the development of U.S. environmental organizations." *Sociological Inquiry* 66:58–83.

———. 2000. *Agency, Democracy, and Nature: The U.S. Environmental Movement from a Critical Theory Perspective.* Cambridge: MIT Press.

Brusick, David J. 1980. *Principles of Genetic Toxicology.* New York: Plenum Press.

———. 1990. "Environmental mutagenesis: An assessment of the past twenty years." In *Mutation and the Environment,* edited by Mortimer L. Mendelson and Richard J. Albertini, 1–16. New York: Wiley-Liss.

Bucher, Rue. 1962. "Pathology: A study of social movements within a profession." *Social Problems* 10:40–51.

Bucher, Rue, and Anselm Strauss. 1961. "Professions in process." *American Journal of Sociology* 66:325–334.

Bullard, Robert D. 1993. *Confronting Environmental Racism: Voices from the Grassroots.* Boston: South End Press.

Cairns, John, Jr. 1968. "We're in hot water." *Scientist and Citizen* 8:187–198.

Cambrosio, Alberto, and Peter Keating. 1983. "The disciplinary stake: The case of chronobiology." *Social Studies of Science* 13:323–353.

Camic, Charles. 1983. *Experience and Enlightenment: Socialization for Cultural Change in Eighteenth-Century Scotland.* Chicago: University of Chicago Press.

Carlson, Elof Axel. 1966. *The Gene: A Critical History.* Philadelphia: W. B. Saunders.

———. 1981. *Genes, Radiation, and Society: The Life and Work of H. J. Muller.* Ithaca, N.Y.: Cornell University Press.

Carson, Rachel. 1962. *Silent Spring.* New York: Fawcett Crest Books.

Chubin, Daryl E. 1976. "The conceptualization of scientific specialties." *Sociological Quarterly* 17:448–476.

Clarke, Adele E. 1998. *Disciplining Reproduction: Modernity, American Life Sciences, and the Problems of Sex.* Berkeley and Los Angeles: University of California Press.

Clarke, Adele E., Janet K. Shim, Laura Mamo, Jennifer Ruth Fosket, and Jennifer R. Fishman. 2003. "Biomedicalization: Technoscientific transformations of health, illness, and U.S. biomedicine." *American Sociological Review* 68:161–194.

Clarke, Adele E., and Joan H. Fujimura, eds. 1992. *The Right Tools for the Job: At Work in Twentieth-Century Life Sciences.* Princeton, N.J.: Princeton University Press.

Cole, Jonathan R., and Harriet Zuckerman. 1975. "The emergence of a scientific specialty: The self-exemplifying case of the sociology of science." In *The Idea of Social Structure: Papers in Honor of Robert K. Merton,* edited by Lewis A. Coser, 140–143. New York: Harcourt Brace Jovanovich.

Collins, Randall. 1998. *The Sociology of Philosophies: A Global Theory of Intellectual Change.* Cambridge: Harvard University Press.

Comfort, Nathaniel C. 2001. *The Tangled Field: Barbara McClintock's Search of the Patterns of Genetic Control.* Cambridge: Harvard University Press.

Corn, Jacqueline Karnell. 1992. *Response to Occupational Hazards: A Historical Perspective.* New York: Van Nostrand Reinhold.

Creager, Angela N. H. 2002. *The Life of a Virus: Tobacco Mosaic Virus as an Experimental Model, 1930–1965.* Chicago: University of Chicago Press.

Crow, James F. 1965. "Genetics and medicine." In *Heritage from Mendel,* edited by R. Alexander Brink, 351–374. Madison: University of Wisconsin Press.

———. 1966. "The evolution of resistance." *Scientist and Citizen* 8:14–19.

———. 1968. "Chemical risk to future generations." *Scientist and Citizen* 10:113–117.

———. 1971a. "The meaning of mutagenicity for society." *EMS Newsletter* 3:22–24.

———. 1971b. "A lethal legacy." In *Science Year 1971,* 94–105. Chicago: Field Enterprises Educational Corporation.

———. 1989. "Concern for environmental mutagens: Some personal reminiscences." *Environmental and Molecular Mutagenesis* 14 (supplement 16):7–10.

Crow, James F., and Seymour Abrahamson. 1965. "Genetic effects of ionization radiation." In *The Science of Ionizing Radiation,* edited by Lewis E. Etter, 263–287. Springfield, Ill.: Charles C. Thomas.

de Chadarevian, Soraya, and Harmke Kamminga, eds. 1998. *Molecular Biology and Medicine: New Practices and Alliances, 1910s–1970s.* Amsterdam: Harwood Academic Publishers.

de Serres, Frederick J. 1981. "Introduction." In *Comparative Chemical Mutagenesis,* edited by Frederick J. de Serres and Michael D. Shelby, 1–3. New York and London: Plenum Press.

de Serres, Frederick J., ed. 1983. *Chemical Mutagens: Principles and Methods for Their Detection.* Vol. 8. New York and London: Plenum Press.

———. 1984. *Chemical Mutagens: Principles and Methods for Their Detection.* Vol. 9. New York and London: Plenum Press.

———. 1986. *Chemical Mutagens: Principles and Methods for Their Detection.* Vol. 10. New York and London: Plenum Press.

de Serres, Frederick J., and Alexander Hollaender, eds. 1980. *Chemical Mutagens: Principles and Methods for Their Detection.* Vol. 6. New York and London: Plenum Press.

———. 1982. *Chemical Mutagens: Principles and Methods for Their Detection*. Vol. 7. New York and London: Plenum Press.

de Serres, Frederick J., and H. G. Kolmark. 1958. "A direct method for determination of forward-mutation rates in *Neurospora crassa*." *Mutation Research* 182:1249–1250.

de Serres, Frederick J., and Michael D. Shelby, eds. 1981. *Comparative Chemical Mutagenesis*. New York: Plenum Press.

Dick, Steven J. 1996. *The Biological Universe: The Twentieth-Century Extraterrestrial Life Debate and the Limits of Science*. New York: Cambridge University Press.

Dickson, David. 1988. *The New Politics of Science*. Chicago: University of Chicago Press.

Divine, Robert A. 1978. *Blowing on the Wind: The Nuclear Test Ban Debate, 1954–1960*. New York: Oxford University Press.

Drake, John W. 1963. "Properties of ultraviolet-induced rII mutations of bacteriophage T4." *Journal of Molecular Biology* 6:268–283.

———. 1970. *The Molecular Basis of Mutation*. San Francisco: Holden-Day.

Drake, John W., and Robert E. Koch. 1976. *Mutagenesis*. Stroudsburg: Dowden, Hutchinson and Ross.

Drake, John W., et al. 1975. "Environmental mutagenic hazards." *Science* 187:503–514.

Dunlap, Riley E. 1992. "Trends in public opinion toward environmental issues: 1965–1990." In *American Environmentalism: The U.S. Environmental Movement, 1970–1990*, edited by Riley E. Dunlap and Angela G. Mertig, 89–116. London: Taylor and Francis.

Echeverria, Ana Barahona. 1995. "Genetic mutation: The development of the concept and its evolutionary implications." In *Mexican Studies in the History and Philosophy of Science*, edited by S. Ramirez and R. S. Cohen, 89–107. Boston: Kluwer Academic Publishers.

Edge, David, and Michael J. Mulkay. 1976. *Astronomy Transformed: The Emergence of Radio Astronomy in Britain*. New York: John Wiley and Sons.

Epstein, Samuel S. 1968. "Testimony on cancer and mutation-producing chemicals in polluted urban air." *U.S. Senate Subcommittee Hearings on Air and Water Pollution of the Committee of Public Works*. July 29–31, Washington, D.C.

———. 1969a. "Chemical mutagens and the Environmental Mutagen Society." *Medical Tribune* 10:11.

———. 1969b. "Chemical mutagens and the Environmental Mutagen Society." *EMS Newsletter* 1:9–11.

———. 1969c. "The role of mutagenicity testing in toxicology." *EMS Newsletter* 2:7–10.

———. 1974. "Introductory remarks to session on 'Mutagens in the Biosphere.'" *Mutation Research* 26:219–223.

———. 1979. *The Politics of Cancer*. New York: Anchor Books.

Epstein, Samuel S., Lester O. Brown, and Carl Pope. 1982. *Hazardous Waste in America*. San Francisco: Sierra Club Books.

Epstein, Samuel S., Joshua Lederberg, Marvin Legator, Arthur H. Wolff, and Alexander Hollaender. 1969. "Cyclamate ban." *Science* 166:1575.

Epstein, Steven. 1996. *Impure Science: AIDS, Activism, and the Politics of Knowledge*. Berkeley and Los Angeles: University of California Press.

Failla, G. 1954. "Editorial." *Radiation Research* 1:1.

Falconer, Douglas. 1993. "Quantitative genetics in Edinburgh: 1947–1980." *Genetics* 133:137–142.

Fishbein, Lawrence. 1973. "Mutagens and potential mutagens in the biosphere." Paper presented at the First International Conference on Environmental Mutagens, August 29–September 1, Asilomar, California.

Food and Drug Administration Advisory Committee on Protocols for Safety Evaluation.

1970. "Panel on Reproduction report on reproduction studies in the safety evaluation of food additives and pesticide residues." *Toxicology and Applied Pharmacology* 16:264–296.

Fortun, Michael, and Everett Mendelsohn, eds. 1999. *The Practices of Human Genetics.* Boston: Kluwer Academic Publishers.

Foucault, Michel. 1978. *The History of Sexuality.* Vol. 1, *An Introduction.* New York: Vintage.

———. 1980. *Power/Knowledge: Selected Interviews and Other Writings 1972–1977.* New York: Random House.

Frickel, Scott. 2004a. "Building an interdiscipline: Collective action framing and the rise of genetic toxicology." *Social Problems* 51:269–287.

———. 2004b. "Scientist activism in environmental justice conflicts: An argument for synergy." *Society and Natural Resources* 17:1–8.

Frickel, Scott, and Debra J. Davidson. 2004. "Building environmental states: Legitimacy and rationalization in sustainability governance." *International Sociology* 19(1):89–110.

Fries, N. 1950. "The production of mutations by caffeine." *Hereditas* 36:134–149.

Fuchs, Stephan, and Peggy S. Plass. 1999. "Sociology and social movements." *Contemporary Sociology* 28:271–277.

Fujimura, Joan H. 1992. "Crafting science: Standardized packages, boundary objects, and 'translation.'" In *Science as Practice and Culture*, edited by Andrew Pickering, 168–211. Chicago: University of Chicago Press.

———. 1996. *Crafting Science: A Sociohistory of the Quest for the Genetics of Cancer.* Cambridge: Harvard University Press.

Galison, Peter. 1996. "Computer simulations and the trading zone." In *The Disunity of Science: Boundaries, Contexts, and Power*, edited by Peter Galison and David J Stump, 118–157. Stanford, Calif.: Stanford University Press.

Geiger, Roger L. 1993. *Research and Relevant Knowledge: American Research Universities since World War II.* New York: Oxford University Press.

Gibbons, Michael, et al. 1994. *The New Production of Knowledge: The Dynamics of Science and Research in Contemporary Societies.* London: Sage.

Gieryn, Thomas F. 1983. "Boundary-work and the demarcation of science from non-science: Strains and interests in professional ideologies of scientists." *American Sociological Review* 48:781–795.

———. 1994. "Boundaries of science." In *Handbook of Science and Technology Studies*, edited by Sheila Jasanoff, Gerald E. Markle, James C. Petersen, and Trevor Pinch, 393–443. Thousand Oaks, Calif.: Sage.

———. 1999. *Cultural Boundaries of Science: Credibility on the Line.* Chicago: Chicago University Press.

Goldstein, Avram. 1962. "Mutagens currently of potential significance to man and other species." In *Mutations, Second Macy Conference on Genetics*, edited by William J. Schull, 167–242. Ann Arbor: University of Michigan Press.

Goodell, Rae S. 1977. *The Visible Scientists.* Boston: Little, Brown.

Gottlieb, Robert. 1993. *Forcing the Spring: The Transformation of the American Environmental Movement.* Washington, D.C.: Island Press.

Gottweis, Herbert. 1998. *Governing Molecules.* Cambridge: MIT Press.

Graham, Frank, Jr. 1970. *Since Silent Spring.* Greenwich, Conn.: Fawcett.

Greenberg, Daniel S. 1999 [1967]. *The Politics of Pure Science.* Chicago: University of Chicago Press.

Griffith, Belver C., and Nicholas C. Mullins. 1972. "Coherent social groups in scientific change." *Science* 177:959–964.

Groenewegen, Peter. 1987. "Attracting audiences and the emergence of toxicology as a practical science." In *The Social Direction of the Public Sciences*, edited by S. Blume, J. Bunders, L. Leydesdorff, and R. Whitley, 307–328. Dordrecht, Boston, and London: D. Reidel

Guillemin, Jeanne. 1999. *Anthrax: The Investigation of a Deadly Outbreak*. Berkeley and Los Angeles: University of California Press.

Gunter, Valerie J., and Craig K. Harris. 1998. "Noisy winter: The DDT controversy in the years before *Silent Spring*." *Rural Sociology* 63:179–198.

Guston, David H. 1999. "Stabilizing the boundary between U.S. politics and science: The rôle of the Office of Technology Transfer as a boundary organization." *Social Studies of Science* 29:87–111.

———. 2000. *Between Science and Politics: Assuring the Integrity and Productivity of Research*. New York: Cambridge University Press.

———, ed. 2001. "Special issue on environmental boundary organizations." *Science, Technology, and Human Values* 26(4).

Hacking, Ian. 1992. "The self-vindication of the laboratory sciences." In *Science as Practice and Culture*, edited by Andrew Pickering, 29–64. Chicago: University of Chicago Press.

Hannigan, John A. 1995. *Environmental Sociology: A Social Constructionist Perspective*. New York: Routledge.

Hayes, Samuel P. 1987. *Beauty, Health, and Permanence: Environmental Politics in the United States, 1955–1985*. New York: Cambridge University Press.

Hays, Harry W. 1986. *Society of Toxicology History, 1961–1986*. Washington, D.C.: Society of Toxicology.

Henry, W. P. 1971. "A beautiful, and poisonous, shade of green." *EMS Newsletter* 4:4.

Hess, David J. Forthcoming. *Science, Technology, and Human Values*.

Hessenbruch, Arne. 2000. "Calibration and work in the X-ray economy, 1896–1928." *Social Studies of Science* 30:397–420.

Hobsbawm, Eric J., and George Rudé. 1968. *Captain Swing: A Social History of the Great English Agricultural Uprising of 1830*. New York: W. W. Norton.

Hoffman, Lily M. 1989. *The Politics of Knowledge: Activist Movements in Medicine and Planning*. Albany: SUNY Press.

Hollaender, A. 1973. "General summary and recommendations for workshop on the evaluation of chemical mutagenicity data in relation to population risk." *Environmental Health Perspectives* 6 (December):229–232.

———, ed. 1971a. *Chemical Mutagens: Principles and Methods for Their Detection*. Vol. 1. New York and London: Plenum Press.

———. 1971b. *Chemical Mutagens: Principles and Methods for Their Detection*. Vol. 2. New York and London: Plenum Press.

———. 1973. *Chemical Mutagens: Principles and Methods for Their Detection*. Vol. 3. New York and London: Plenum Press.

———. 1976. *Chemical Mutagens: Principles and Methods for Their Detection*. Vol. 4. New York and London: Plenum Press.

Hollaender, Alexander, and Frederick J. de Serres, eds. 1978. *Chemical Mutagens: Principles and Methods for Their Detection*. Vol. 5. New York: Plenum Press.

Hughes, Tom. 1999. "Symposium: Watershed and drinking water toxicology: Studies at EPA." *EMS Newsletter* (summer):7–9.

Inglehart, Ronald. 1990. *Culture Shift in Advanced Industrial Society*. Princeton, N.J.: Princeton University Press.

Johnson, Leland, and Daniel Schaffer. 1994. *Oak Ridge National Laboratory: The First Fifty Years.* Knoxville: University of Tennessee Press.

Johnston, Ron, and Dave Robbins. 1977. "The development of specialties in industrialized science." *Sociological Review* 25:87–108.

Kay, Lilly E. 1993. *The Molecular Vision of Life: Caltech, the Rockefeller Foundation, and the Rise of the New Biology.* New York: Oxford University Press.

———. 2000. *Who Wrote the Book of Life? A History of the Genetic Code.* Stanford, Calif.: Stanford University Press.

Keating, Terry J. 2001. "Lessons from the recent history of the Health Effects Institute." *Science, Technology, and Human Values* 26:409–430.

Keller, Evelyn Fox. 1983. *A Feeling for the Organism: The Life and Work of Barbara McClintock.* New York: Freeman.

Kennedy, Michael. 1990. "The constitution of critical intellectuals: Polish physicians, peace activists and democratic civil society." *Studies in Comparative Communism* 23:281–303.

Kenney, Martin. 1986. *Biotechnology: The University-Industrial Complex.* New Haven, Conn.: Yale University Press.

Kevles, Daniel J. 1978. *The Physicists: The History of a Scientific Community in Modern America.* New York: Alfred A. Knopf.

———. 1985. *In the Name of Eugenics: Genetics and the Uses of Human Heredity.* Berkeley and Los Angeles: University of California Press.

Kihlman, B. A. 1964. "The production of chromosomal aberrations by streptonogrin in *vicia faba*." *Mutation Research* 1:54–62.

Kilbey, Brian J. 1995. "In memoriam Charlotte Auerbach, FRS (1899–1994)." *Mutation Research* 327:1–4.

Kimmelman, Barbara Ann. 1987. "A Progressive Era discipline: genetics at American agricultural colleges and experiment stations, 1900–1920." Ph.D. diss., University of Pennsylvania.

Kinchy, Abby J., and Daniel Lee Kleinman. 2003. "Organizing credibility: Discursive and organizational orthodoxy on the borders of ecology and politics." *Social Studies of Science* 33(4):1–28.

———. Forthcoming. "On the borders of post-war ecology: organizational and discursive legitimacy and the construction of scientific boundaries." *Science, Technology and Human Values.*

Kitschelt, Herbert P. 1986. "Political opportunity structures and political protest: Anti-war movements in four democracies." *British Journal of Political Science* 16:57–85.

Klein, Julie Thompson. 1996. *Crossing Boundaries: Knowledge, Disciplinarities, and Interdisciplinarities.* Charlottesville: University Press of Virginia.

Kleinman, Daniel L. 1995. *Politics on the Endless Frontier: Postwar Research Policy in the United States.* Durham, N.C.: Duke University Press.

———. 2003. *Impure Cultures: University Biology and the World of Commerce.* Madison: University of Wisconsin Press.

———, ed. 2000. *Science, Technology, and Democracy.* Albany: SUNY Press.

Kleinman, Daniel L., and Mark Solovey. 1995. "Hot science/cold war: The National Science Foundation after World War II." *Radical History Journal* 63:110–139.

Kleinman, Daniel L., and Steven P. Vallas. 2001. "Science, capitalism, and the rise of the 'knowledge worker': The changing structure of knowledge production in the United States." *Theory and Society* 30:451–492.

Knipling, E. F. 1962. "Potentialities and progress in the development of chemosterilants for insect control." *Journal of Economic Entomology* 55:782–786.

Knorr, Karin. 1977. "Producing and reproducing new knowledge: Descriptive or constructive? Toward a model of research production." *Social Science Information* 16:669–696.

Kohler, Robert E. 1982. *From Medical Chemistry to Biochemistry: The Making of a Biomedical Discipline*. Cambridge: Cambridge University Press.

———. 1991. *Partners in Science: Foundations and Natural Scientists, 1900–1945*. Chicago: University of Chicago Press.

———. 1994. *Lords of the Fly:* Drosophila *Genetics and the Experimental Life*. Chicago: University of Chicago Press.

Krimsky, Sheldon. 2000. *Hormonal Chaos: The Scientific and Social Origins of the Environmental Endocrine Hypothesis*. Baltimore: Johns Hopkins University Press.

Krohn, Wolfgang, and Wolf Schafer. 1976. "The origins and structure of agricultural chemistry." In *Perspectives on the Emergence of Scientific Disciplines*, edited by Gerard Lemaine, Roy Macleod, Michael Mulkay, and Peter Weingart, 27–52. Chicago: Aldine.

Kroll-Smith, Steve, Phil Brown, and Valerie J. Gunter, eds. 2000. *Environment and Illness: A Reader in Contested Medicine*. New York: New York University Press.

Kuznick, Peter J. 1987. *Beyond the Laboratory: Scientists as Political Activists in 1930s America*. Chicago: University of Chicago Press.

Kwa, Chunglin. 1987. "Representations of nature mediating between ecology and science policy: The case of the International Biological Programme." *Social Studies of Science* 17:413–442.

LaBreque, G. C., P. H. Adcock, and C. N. Smith. 1960. "Tests with compounds effecting house fly metabolism." *Journal of Economic Entomology* 53:802.

Larson, Magali Sarfatti. 1984. "The production of expertise and the constitution of expert power." In *The Authority of Experts*, edited by Thomas Haskell, 28–80. Bloomington: University of Indiana Press.

Latour, Bruno. 1987. *Science in Action*. Cambridge: Harvard University Press.

———. 1988. *The Pasteurization of France*. Cambridge: Harvard University Press.

———. 1993. *We Have Never Been Modern*. Cambridge: Harvard University Press.

Latour, Bruno, and Steve Woolgar. 1986. *Laboratory Life: The Construction of Scientific Facts*. Princeton, N.J.: Princeton University Press.

Lattuca, Lisa R. 2001. *Creating Interdisciplinarity: Interdisciplinary Research and Teaching among College and University Faculty*. Nashville, Tenn.: Vanderbilt University Press.

Law, John. 1976. "The development of specialties in science: The case of x-ray protein crystallography." In *Perspectives on the Emergence of Scientific Disciplines*, edited by Gerard Lemaine, Roy MacLoed, Michael Mulkay, and Peter Weingart, 123–152. Chicago: Aldine.

Layton, Edward, Jr. 1971. *The Revolt of the Engineers: Social Responsibility and the American Engineering Profession*. Cleveland: Case Western Reserve University Press.

Lederberg, Joshua. 1955. Letter to the editor. *Bulletin of the Atomic Scientists* 11:365.

———. 1963a. "Biological future of man." In *Man and His Future*, edited by Gordon Wolstenholme, 263–273. Boston: Little, Brown.

———. 1963b. "Molecular biology, eugenics, and euphenics." *Nature* 198:428–429.

———. 1967. "A test tube daddy." *Washington Post*, April 16.

———. 1969. "Environmental chemicals' hazards still little known." *Washington Post*, November 1.

———. 1970. Preface to *Drugs of Abuse: Their Genetic and Other Psychiatric Hazards*, edited by Samuel S. Epstein. Washington, D.C.: U.S. GPO.

———. 1971. "Food Additives." In *The Social Responsibility of the Scientist*, edited by Martin Brown, 121–130. New York: Free Press.

———. 1997. "Some early stirrings (1950 ff.) of concern about environmental mutagens." *Environmental and Molecular Mutagenesis* 30:3–10.

Legator, Marvin S. 1970. "Chemical mutagenesis comes of age: Environmental implications." *Journal of Heredity* 61:239–242.

Legator, Marvin S., and Samuel S Epstein. 1970. "Reply to Dr. Zbinden's letter." *EMS Newsletter* 1(3):4.

Legator, Marvin S., Barbara L. Harper, and Michael J. Scott. 1985. *The Health Detective's Handbook: A Guide to the Investigation of Environmental Health Hazards by Nonprofessionals.* Baltimore: Johns Hopkins University Press.

Legator, Marvin S., and Heinrich V. Malling. 1971. "The host mediated assay, a practical procedure for evaluating potential mutagenic agents in mammals." In *Chemical Mutagens: Principles and Methods for Their Detection*, vol. 2, edited by Alexander Hollaender, 569–590. New York: Plenum Press.

Lemaine, Gerard, Roy MacLoed, Michael Mulkay, and Peter Weingart, eds. 1976. *Perspectives on the Emergence of Scientific Disciplines.* Chicago: Aldine.

Lenoir, Timothy. 1997. *Instituting Science: The Cultural Production of Scientific Disciplines.* Stanford, Calif.: Stanford University Press.

Levitt, Norman, and Paul R. Gross. 1994. *Higher Superstition: The Academic Left and Its Quarrels with Science.* Baltimore: Johns Hopkins University Press.

Lindee, M. Susan. 1992. "What is a mutation? Identifying heritable change in the offspring of survivors at Hiroshima and Nagasaki." *Journal of the History of Biology* 25:231–255.

———. 1994. *Suffering Made Real: American Science and the Survivors at Hiroshima.* Chicago: University of Chicago Press.

———. Forthcoming. *Moments of Truth: Genetic Disease in American Culture.* Baltimore: Johns Hopkins University Press.

Loveless, Anthony. 1966. *Genetic and Allied Effects of Alkylating Agents.* University Park: Pennsylvania State University Press.

Ma, Te-Hsiu. 1995. "Trandescantia (spiderwort) plants as biomonitors of the genotoxicity of environmental pollutants." In *Biomonitors and Biomarkers as Indicators of Environmental Change*, edited by F. M. Butterworth et al., 207–216. New York: Plenum Press.

Malling, Heinrich V. 1970. "Chemical mutagens as a possible genetic hazard in human populations." *Journal of the American Hygiene Association* 31:657–666.

———. 1971. "Environmental Mutagen Information Center (EMIC) II. Development for the Future." *EMS Newsletter* 4 (March):11–15.

———. 1977. "Goals and programs of the Laboratory of Environmental Mutagenesis." *Environmental Health Perspectives* 20:263–265.

Malling, H. V., and J. S. Wassom. 1969. "Environmental Mutagen Information Center (EMIC) I. Initial organization." *EMS Newsletter* 1(1):16–18.

Malling, H. V., J. S. Wassom, and S. S. Epstein. 1970. "Mercury in our environment." *EMS Newsletter* 1(3):7–9.

Markowitz, Gerald, and David Rosner. 2002. *Deceit and Denial: The Deadly Politics of Industrial Pollution.* Berkeley and Los Angeles: University of California Press.

Martin, Emily. 1998. "Anthropology and the cultural study of science." *Science, Technology, and Human Values* 23:24–44.

Mazur, Allen, and J. Lee. 1993. "Sounding the global alarm: Environmental issues in the U.S. national news." *Social Studies of Science* 23:681–720.

McAdam, Doug, John D. McCarthy, and Mayer N. Zald, eds. 1996. *Comparative Perspectives on Social Movements.* New York: Cambridge University Press.

McAdam, Doug, Sidney Tarrow, and Charles Tilly. 2001. *Dynamics of Contention.* New York: Cambridge University Press.

McCally, Michael. 2002. "Medical activism in environmental health." *Annals of the American Academy of Political and Social Science* 584:145–158.

McGucken, William. 1984. *Scientists, Society, and the State: The Social Relations of Science Movement in Great Britain*, 1931–1947. Columbus: Ohio State University Press.

McKibben, Bill. 1990. *The End of Nature.* New York: Anchor Books.

Mertig, Angela G., Riley E. Dunlap, and Denton E. Morrison. 2002. "The environmental movement in the United States." In *Handbook of Environmental Sociology*, edited by Riley E. Dunlap and William Michelson, 448–481. Westport, Conn.: Greenwood Press.

Merton, Robert K. 1973. *The Sociology of Science.* Chicago: University of Chicago Press.

Meselson, Matthew. 1971. Preface to *Chemical Mutagens: Principles and Methods for Their Detection*, vol. 1, edited by Alexander Hollaender, ix–xii. New York: Plenum Press.

Messer-Davidow, Ellen, David R. Shumway, and David J. Sylvan 1993. "Disciplinary ways of knowing." In Knowledges: *Historical and Critical Studies in Disciplinarity*, edited by Ellen Messer-David, David R. Shumway, and David J. Sylvan, 1–21. Charlottesville: University Press of Virginia.

——, eds. 1993. *Knowledges: Historical and Critical Studies in Disciplinarity.* Charlottesville: University Press of Virginia.

Meyer, David S., and Suzanne Staggenborg. 1996. "Movements, countermovements, and the structure of political opportunity." *American Journal of Sociology* 101:1628–1660.

Miller, Clark A. 2001. "Hybrid Management: Boundary organizations, science policy, and environmental governance in the climate regime." *Science, Technology, and Human Values* 26:478–500.

Miller, Clark A., and Paul N. Edwards, eds. 2001. *Changing the Atmosphere: Expert Knowledge and Environmental Governance.* Cambridge: MIT Press.

Mitchell, Robert Cameron, Angela G. Mertig, and Riley E. Dunlap. 1992. "Twenty years of environmental mobilization: Trends among national environmental organizations." In *American Environmentalism: The U.S. Environmental Movement, 1970–1990*, edited by Riley E. Dunlap and Angela G. Mertig, 11–26. Philadelphia: Taylor and Francis.

Moore, Kelly. 1996. "Organizing integrity: American science and the creation of public interest science organizations, 1955–1975." *American Journal of Sociology* 101: 1592–1627.

——. Forthcoming. *Disruptive Science: Professionals, Activism, and the Politics of War in the United States, 1945–1975.* Princeton, N.J.: Princeton University Press.

Moran, Joe. 2002. *Interdisciplinarity.* London and New York: Routledge.

Morange, Michel. 2000. *A History of Molecular Biology.* Cambridge: Harvard University Press.

Morris, Aldon D., and Carol McClurg Mueller, eds. 1992. *Frontiers in Social Movement Theory.* New Haven, Conn.: Yale University Press.

Morris, Jim. 1997. "New alarm over hydrogen sulfide: Researchers document lasting damage to human nervous system." *Houston Chronicle*, November 13.

Mulkay, Michael. 1979. *Science and the Sociology of Knowledge.* London: Allen and Unwin.

Muller, H. J. 1927. "Artificial transmutation of the gene." *Science* 66:84–87.

——. 1935. *Out of the Night: A Biologist's View of the Future.* New York: Vanguard.

——. 1950. "Our load of mutations." *American Journal of Human Genetics* 2:111–176.

——. 1955a. "How radiation changes the genetic constitution." *Bulletin of the Atomic Scientists* 11:329–338.

——. 1955b. "Radiation and human mutation." *Scientific American* 195:58–66.

——. 1962a. "Artificial transmutation of the gene." In *Studies in Genetics: The Selected Papers of H. J. Muller*, 245–251. Bloomington: University of Indiana Press.

——. 1962b. *Studies in Genetics: The Selected Papers of H. J. Muller.* Bloomington: University of Indiana Press.

———. 1962c. "The problem of genic modification." In *Studies in Genetics: The Selected Papers of H. J. Muller*, 252–276. Bloomington: University of Indiana Press.

———. 1963. "Genetic progress by voluntarily conducted germinal choice." In *Man and His Future*, edited by Gordon Wolstenholme, 247–262. Boston: Little, Brown.

———. 1965. "Means and aims in human genetic betterment." In *The Control of Human Heredity and Evolution*, edited by T. M. Sonneborn, 100–122. New York: Macmillan.

Mullins, Nicholas C. 1973. *Theories and Theory Groups in Contemporary American Sociology.* New York: Harper and Row.

———. 1976. "The development of a scientific specialty: The Phage Group and the origins of molecular biology." *Minerva* 10:51–82.

Muskie, Edmund S. 1969. "Chemicals, the toxicologist, and the future of man." *Forum for the Advancement of Toxicology* 2:1, 3.

National Academy of Sciences Agricultural Board. 1961. *Symposium on Mutation and Plant Breeding.* Washington, D.C.: National Academy of Sciences–National Research Council.

National Academy of Sciences–National Research Council. 1956. *The Biological Effects of Atomic Radiation. Summary Reports.* Washington, D.C.: National Academy of Science–National Research Council.

———. 1960. *The Biological Effects of Atomic Radiation. Summary Reports from a Study by the National Academy of Sciences.* Washington, D.C.: National Academy of Sciences–National Research Council.

———. 1972. *The Effects on Populations of Exposure to Low Levels of Ionizing Radiations.* Washington, D.C.: National Academy of Sciences–National Research Council.

National Center for Toxicogenomics. 2003. *Using Global Genomic Expression Technology to Create a Knowledge Base of Protecting Human Health* (http://www.niehs.nih.gov/nct/).

National Institute of Environmental Health Sciences (NIEHS). 1972. *Annual Report.* Research Triangle Park, N.C.: National Institute of Environmental Health Sciences.

———. 1975a. *Annual Report.* Research Triangle Park, N.C.: National Institute of Environmental Health Sciences.

———. 1975b. *Data Book: Permanent Facility.* Research Triangle Park, N.C.: National Institute of Environmental Health Sciences.

———. 1977. *Annual Report.* Research Triangle Park, N.C.: National Institute of Environmental Health Sciences.

Neel, James V. 1970. "Lessons from a 'primitive' people: Do recent data concerning South American Indians have relevance to problems of highly civilized communities?" *Science* 170:815–822.

Neel, James V., and Arthur D. Bloom. 1969. "The detection of environmental mutagens." *Medical Clinics of North America* 53:1243–1256.

Neel, James V., and William J. Schull. 1956. *The Effect of Exposure to the Atomic Bombs on Progeny Termination in Hiroshima and Nagasaki.* Washington, D.C.: National Academy of Sciences.

Nelkin, Dorothy. 1977. "Scientists and professional responsibility: The experience of American ecologists." *Social Studies of Science* 7:75–95.

Nilan, R. A. 1973. "Increasing the effectiveness, efficiency and specificity of mutation induction in flowering plants." In *Genes, Enzymes, and Populations*, edited by A. Srb, 205–222. New York: Plenum Press.

Nowotny, Helga, and Hilary Rose, eds. 1979. *Counter-Movements in the Sciences.* Dordrecht, Boston, and London: D. Reidel.

Nyhart, Lynn K. 1995. *Biology Takes Form: Animal Morphology and the German Universities, 1800–1900.* Chicago: University of Chicago Press.

Oak Ridge National Laboratory Biology Division. June 1968. *Annual Progress Report for 1967.* Oak Ridge, Tenn.: Oak Ridge National Laboratory Biology Division.

———. 1969. *Annual Progress Report for 1968.* Oak Ridge, Tenn.: Oak Ridge National Laboratory Biology Division.

———. 1970. *Annual Progress Report for 1969.* Oak Ridge, Tenn. Oak Ridge National Laboratory Biology Division.

Oliver, Pamela E., and Daniel J. Myers. 1999. "How events enter the public sphere: Conflict, location, and sponsorship in local newspaper coverage of public events." *American Journal of Sociology* 105:38–87.

Palladino, Paulo. 1996. *Entomology, Ecology and Agriculture: The Making of Scientific Careers in North America 1885–1985.* Amsterdam: Harwood Academic Publishers.

Pandora, Katherine. 1997. *Rebels within the Ranks: Psychologists' Critique of Scientific Authority and Democratic Realities in New Deal America.* New York: Cambridge University Press.

Paul, Diane B. 1984. "Eugenics and the Left." Journal of the History of Ideas 45:578–581.

———. 1987. "'Our Load of Mutations' revisited." *Journal of the History of Biology* 20:321–335.

———. 1992. "Eugenic anxieties, social realities, and political choices." *Social Research* 59:663–683.

———. 1995. *Controlling Human Heredity, 1865 to the Present.* Atlantic Highlands, N.J.: Humanities Press.

Pauly, Philip J. 1987. *Controlling Life: Jacques Loeb and the Engineering Ideal in Biology.* New York: Oxford University Press.

———. 2000. *Biologists and the Promise of American Life.* Princeton, N.J.: Princeton University Press.

Peterson, Roger Tory. 1934. *A Field Guide to the Birds; Giving Field Marks of All Species Found in Eastern North America.* Boston: Houghton Mifflin.

Pickering, Andrew. 1993. "The mangle of practice: Agency and emergence in the sociology of science." *American Journal of Sociology* 99:559–589.

Potter, Harry R. 1997. "Precursors of the environmental movement II: Scientists and the legislation of the 1960's." Paper presented at the annual meeting of the American Sociological Association, August 9–13, Toronto.

Powell, Walter W., and Paul DiMaggio, eds. 1991. *The New Institutionalism and Organizational Analysis.* Chicago: University of Chicago Press.

Preston, Julian R., and George R. Hoffman. 2001. "Genetic toxicology." In *Casarett and Doull's Toxicology: The Basic Science of Poisons*, edited by Curtis D. Klaassen, 321–350. New York: McGraw-Hill.

Price, Derek J. de Solla. 1963. *Little Science, Big Science.* New York: Columbia University Press.

Primack, Joel, and Frank von Hippel. 1974. *Advice and Dissent: Scientists in the Political Arena.* New York: Basic Books.

Prival, Michael J., and Vicki L. Dellarco. 1989. "Evolution of social concerns and environmental policies for chemical mutagens." *Environmental and Molecular Mutagenesis* 14 (supplement 16):46–50.

Proctor, Robert N. 1995. *Cancer Wars: How Politics Shapes What We Know and Don't Know About Cancer.* New York: Basic Books.

Research Triangle Institute. 1965. "Recommendations for the Development and Operation of the National Environmental Health Sciences Center." Research Triangle Park, N.C.: Department of Health, Education, and Welfare, U.S. Public Health Service, Bureau of State Services (Environmental Health).

Rosenberg, Charles E. 1979. "Toward an ecology of knowledge: On discipline, context, and history." In *The Organization of Knowledge in Modern America, 1860–1920*, edited by Alexandra Oleson and John Voss, 440–455. Baltimore: Johns Hopkins University Press.

———. 1997. *No Other Gods: On Science and American Social Thought*. Baltimore: Johns Hopkins University Press.

Rosner, David, and Gerald Markowitz. 1987. *Dying for Work: Workers' Safety and Health in Twentieth-Century America*. Bloomington: University of Indiana Press.

Rotblat, Joseph. 1972. *Scientists in the Quest for Peace*. Cambridge: MIT Press.

Russell, Liane B. 1994. "Role of mouse germ-cell mutagenesis in understanding genetic risk and in generating mutations that are prime tools for studies in modern biology." *Environmental and Molecular Mutagenesis* 23:23–29.

Russell, William L. 1989. "Reminiscences of a mouse specific-locus addict." *Environmental and Molecular Mutagenesis* 14:16–22.

Sale, Kirkpatrick. 1993. *The Green Revolution: The American Environmental Movement, 1962–1992*. New York: Hill and Wang.

Samson, Leona D. 2003. "On the 50th anniversary of solving the structure of DNA." *Environmental Health Perspectives* 111:A329–A331.

Sanders, Howard J. 1969a. "Chemical mutagens: The road to genetic disaster?" *Chemical and Engineering News* 47 (May 19):50–65, 71.

———. 1969b. "Chemical mutagens: An expanding roster of suspects." *Chemical and Engineering News* 47 (June 2):54–68.

Sanders, Norman K. 1972. *Stop It! A Guide to Defense of the Environment*. San Francisco: Rinehart Press.

Sankaranarayanan, K., and P.H.M. Lohman. 1993. "In memoriam Frederik Hendrik Sobels (February 22, 1922–July 6, 1993)." *Mutation Research* 29:102–104.

Sapp, Jan. 1987. *Beyond the Gene: Cytoplasmic Inheritance and the Struggle for Authority in Genetics*. New York: Oxford University Press.

Sarewitz, Daniel. 1996. *Frontiers of Illusion: Science, Technology, and the Politics of Progress*. Philadelphia: Temple University Press.

Schmeck, Harold M., Jr. 1970a. "Genetic defects may be screened." *New York Times*, November 6.

———. 1970b. "Mutation linked to 'gas' additive." *New York Times*, November 7.

Schull, William J., ed. 1962. *Mutations, Second Macy Conference on Genetics*. Ann Arbor: University of Michigan Press.

Scott, D., and H. J. Evans. 1964. "On the non-requirement for deoxyribonucleic acid synthesis in the production of chromosome aberrations by 8-ethoxycaffeine." *Mutation Research* 1:146–156.

Sellers, Christopher C. 1997. *Hazards of the Job: From Industrial Disease to Environmental Health Science*. Chapel Hill: University of North Carolina Press.

Setlow, Richard B. 1968. "Alexander Hollaender." *Photochemistry and Photobiology* 8:511–513.

———. 1987. "Alexander Hollaender (1898–1986)." *Genetics* 116:1–3.

Shostak, Sara. 2003a. "Locating gene-environment interaction: At the intersections of genetics and public health." *Social Science and Medicine* 56:2327–2342.

———. 2003b. "Disciplinary emergence in the environmental health sciences, 1950–2000." Ph.D. diss., University of California–San Francisco.

Small, Mario L. 1999. "Departmental conditions and the emergence of new disciplines: Two cases in the legitimation of African-American Studies." *Theory and Society* 28:659–707.

Smith, Alice Kimball. 1965. *A Peril and a Hope: The Scientists' Movement in America, 1945–47*. Chicago: University of Chicago Press.

Smith, Roger H. 1971. "Autocide instead of chemical insecticides." *EMS Newsletter* 5:5.

Snow, David A., E. Burke Rochford Jr., Steven K. Worden, and Robert D. Benford. 1986. "Frame alignment processes, micromobilization, and movement participation." *American Sociological Review* 51:464–481.

Snow, David A., and Robert D. Benford. 1988. "Ideology, frame resonance, and participant mobilization." In *From Structure to Action: Comparing Social Movement Research Across Cultures*, edited by B. Klandermans, H. Kriesi, and S. Tarrow, 197–218. Greenwich, Conn.: JAI Press.

———. 1992. "Master frames and cycles of protest." In *Frontiers in Social Movement Theory*, edited by Aldon D. Morris and Carol McClurg Muller, 133–155. New Haven, Conn.: Yale University Press.

Sobels, F. H. 1964. "Preface." *Mutation Research* 1:1.

———. 1975. "Charlotte Auerbach and chemical mutagenesis." *Mutation Research* 29:171–180.

Sonneborn, T. M., ed. 1965. *The Control of Human Heredity and Evolution.* New York: Macmillan.

Sparrow, A. H., and L. A. Schairer. 1971. "Mutational response in *Tradescantia* after accidental exposure to a chemical mutagen." *EMS Newsletter* 1:16–19.

Star, Susan Leigh. 1989. *Regions of the Mind: Brain Research and the Quest for Scientific Certainty.* Stanford, Calif.: Stanford University Press.

Star, Susan Leigh, and James Griesemer. 1989. "Institutional ecology, 'translations' and boundary objects: Amateurs and professionals in Berkeley's museum of vertebrate zoology, 1907–1939." *Social Studies of Science* 12:129–140.

Straney, Sister Margaret J., and Thomas R Mertens. 1969. "A survey of introductory college genetics courses." *Journal of Heredity* 60:223–228.

Tarrow, Sidney. 1989. *Struggle, Politics, and Reform: Collective Action, Social Movements, and Cycles of Protest.* Ithaca, N.Y.: Cornell University Press.

———. 1992. "Mentalities, political cultures, and collective action frames: Constructing meanings through action." In *Frontiers in Social Movement Theory*, edited by Aldon D. Morris and Carol McClurg Mueller, 174–202. New Haven, Conn.: Yale University Press.

———. 1994. *Power in Movement: Social Movements, Collective Action, and Politics.* New York: Cambridge University Press.

———. 1996. "States and opportunities: The political structuring of social movements." In *Comparative Perspectives on Social Movements*, edited by Doug McAdam, John D. McCarthy, and Mayer N. Zald, 41–61. New York: Cambridge University Press.

———. 1998. *Power in Movement: Social Movements and Contentious Politics.* 2nd ed. New York: Cambridge University Press.

Tennant, Raymond W., et al. 1987. "Prediction of chemical carcinogenicity in rodents from in vitro genetic toxicity assays." *Science* 236:933–941.

Tesh, Sylvia N. 2000. *Uncertain Hazards: Environmental Activists and Scientific Proof.* Ithaca, N.Y.: Cornell University Press.

Thackray, Arnold, ed. 1998. *Private Science: Biotechnology and the Rise of the Molecular Sciences.* Philadelphia: University of Pennsylvania Press.

Thackray, Arnold, and Robert K. Merton. 1972. "On discipline building: The paradoxes of George Sarton." *Isis* 63:473–475.

Tilly, Charles. 1978. From *Mobilization to Revolution.* Reading, Mass.: Addison-Wesley.

Turner, James S. 1970. *The Chemical Feast.* New York: Grossman.

Turner, Stephen. 2000. "What are disciplines? And how is interdisciplinarity different?" In *Practicing Interdisciplinarity*, edited by Peter Weingart and Nico Stehr, 46–65. Toronto: University of Toronto Press.

Underbrink, A. G., L. A. Schairer, and A. H Sparrow. 1973. "Tradescantia stamen hairs: A radiobiological test system applicable to chemical mutagensis." In *Chemical Mutagens: Principles and Methods for Their Detection*, vol. 3, edited by Alexander Hollaender, 171–207. New York: Plenum Press.

United Nations. 1958. "Report of the United Nations Scientific Committee on the Effects of Atomic Radiation." Official Records of the General Assembly, Seventeenth Session, Supplement No. 17 (A/3838).

U.S. Congress. 1969. Congressional Record. 91st Congress, First Session. Washington, D.C.: U.S. GPO.

U.S. Department of Health, Education, and Welfare. 1969. *Report of the Secretary's Commission on Pesticides and Their Relationship to Environmental Health.* Washington, D.C.: U.S. GPO.

U.S. Senate. 1971. *Chemicals and the Future of Man: Hearings Before the Subcommittee on Executive Reorganization and Government Research.* Washington, D.C.: U.S. GPO.

van den Bosch, Robert, and P. S. Messenger. 1973. *Biological Control.* New York: Intext Press.

von Borstel, R. C., and Charles M. Steinberg. 1996. "Alexander Hollaender: Myth and mensch." *Genetics* 143:1051–1056.

Wallace, Bruce, and Joseph O. Falkinham. 1997. *The Study of Gene Action.* Ithaca, N.Y.: Cornell University Press.

Wallace, Robert W. 1989. "The institutionalization of a new discipline: The case of sociology at Columbia University, 1891–1931." Ph.D. diss., Columbia University.

Walters, Ronald G., ed. 1997. *Scientific Authority and Twentieth-Century America.* Baltimore: Johns Hopkins University Press.

Walton, John. 1992. "Making the theoretical case." In *What Is a Case?*, edited by Charles C. Ragin and Howard S. Becker, 121–37. New York: Cambridge University Press.

Wang, Jessica. 1999. *American Science in an Age of Anxiety: Scientists, Anti-Communism, and the Cold War.* Chapel Hill: University of North Carolina Press.

Wassom, John. 1973. "The literature on chemical mutagenesis." In *Chemical Mutagens: Principles and Methods for Their Detection*, vol. 3, edited by Alexander Hollaender, 271–287. New York: Plenum Press.

———. 1989. "Origins of genetic toxicology and the Environmental Mutagen Society." *Environmental and Molecular Mutagenesis* 14 (supplement 16):1–6.

Waters, Michael D. 1979. "The Gene-Tox program." In *Mammalian Cell Mutagenesis: The Maturation of Test Systems*, edited by Abraham W. Hsie, Patrick J. O'Neil, and Victor K. McElheny, 449–457. Long Island, N.Y.: Cold Springs Harbor Laboratory.

Weingart, Peter, and Nico Stehr. 2000. "Introduction" to *Practicing Interdisciplinarity*, edited by Peter Weingart and Nico Stehr, xi–xvi. Toronto: University of Toronto Press.

———, eds. 2000. *Practicing Interdisciplinarity.* Toronto: University of Toronto Press.

Whitley, Richard. 1974. "Cognitive and social institutionalization of scientific specialties and research areas." In *Social Processes of Scientific Development*, edited by R. Whitley, 69–95. London: Routledge and Keegan Paul.

———. 1976. "Umbrella and polytheistic scientific disciplines and their elites." *Social Studies of Science* 6:471–497.

———. 1984. *The Intellectual and Social Organization of the Sciences.* Oxford: Clarendon.

Whorton, James. 1975. *Before Silent Spring: Pesticides and Public Health in Pre-DDT America.* Princeton, N.J.: Princeton University Press.

Winner, Langdon. 1977. *Autonomous Technology: Technics-Out-of-Control As a Theme in Political Thought.* Cambridge: MIT Press.

Wit, S. L., P. A. Greve, and A. W. Fonds. 1970. "Endosulfan in the Rhine." *EMS Newsletter* 3:5.

Wolfe, Audra J. 2002. "Speaking for nature and nation: Biologists as public intellectuals in Cold War culture." Ph.D. diss., University of Pennsylvania.

Woodhouse, Edward J., and Steve Breyman. Forthcoming. "Green chemistry as an expert social movement?" *Science, Technology, and Human Values.*

Worster, Donald. 1994. *Nature's Economy: A History of Ecological Ideas.* New York: Cambridge University Press.

Wright, J. W., and R. Pal, eds. 1967. *Genetics of Insect Vectors of Disease.* New York: Elsevier.

Wright, Susan. 1994. *Molecular Politics: Developing American and British Regulatory Policy for Genetic Engineering, 1972–1982.* Chicago: University of Chicago Press.

Zald, Mayer N., and Michael A. Berger. 1978. "Social movements in organizations: Coup d'etat, insurgency, and mass movements." *American Journal of Sociology* 83:823–861.

Zbinden, Gerhard. 1970. "A word of caution." *EMS Newsletter* 1(3):4.

INDEX

ABOUT THE AUTHOR

SCOTT FRICKEL is an assistant professor of sociology at Tulane University in New Orleans, Louisiana, where he teaches courses on science and technology studies, environmental sociology, and social movements. He has published several articles in scholarly journals and is co-editor of *The New Political Sociology of Science: Institutions, Networks, and Power* (forthcoming). *Chemical Consequences* is his first book.